Systems Theory for Engineering Practice

Maosheng Zheng · Jie Yu

Systems Theory for Engineering Practice

Insights from Physics

Springer

Maosheng Zheng
School of Chemical Engineering
Northwest University
Xi'an, Shaanxi, China

Jie Yu
School of Life Science
Northwest University
Xi'an, Shaanxi, China

ISBN 978-981-97-9341-9   ISBN 978-981-97-9342-6 (eBook)
https://doi.org/10.1007/978-981-97-9342-6

This Springer imprint is published by the registered company Springer Nature Singapore Pte Ltd.
The registered company address is: 152 Beach Road, #21-01/04 Gateway East, Singapore 189721, Singapore

Preface

This book presents cognition of the universality of systems theory thinking by using some ordinary physical phenomena and relevant analysis, of which the methodologies are consistent with the viewpoint of systems theory. An involuntary question is that what is "systems science", of course it is the "science about systems". While the term "system" refers to an "organic whole" with specific function or role, which composes of several "constituent elements". These elements interact and depend on each other to form a system. A system as an organic whole can be a part of a larger system as well. So, in principle, it is rather complex in studying "systems science" in general speaking. Till now "systems science" cannot be considered as a mature subject. Fortunately, modeling is a usual methodology for recognizing such things, therefore "systems science" can be acquainted and described by using appropriate models and modeling in considering the wholeness and interactions among its elements of the specific system in a sense. Modeling can play a role of analogy to get knowledge, which could furthermore activate wisdom through overall analyzing information comprehensively.

As to model, famous statistician Box's once stated that "*all models are wrong, but some are useful*". This can be seen as a convenient routine to measure the effectiveness of a modeling. A key insight is to take the modeling process and the relevant predictions as a "way of thinking" to recognize the complex system in some respects. One model is just a model, which is only an expression of understanding reality from analogical or abstractive measurements in some respects; of course it is not the object entity or even an actual representation of the reality. On the other hand, one's understanding and expression in words are always limited than the real complexity of actual thing, so it could not expect any model to fully reveal the reality of whole thing in this point. "A blind feeling an elephant" is a very common thing, only some partial apparent characteristics could be felt instead of the reality itself, but understanding and recognition could be deepened gradually. Great physicist Galileo said, "measure everything that can be measured, and turn the un-measurable into measurable", so scientific research is permanently exploring the "measurement" work of everything in the world.

Systems science is not in exception, we could enumerate some actual examples from natural world, which have the features of "systems science" and are characterized by appropriate models and modeling gradually. These descriptions reflect the universality of systems theory thinking in ordinarily physical methodologies.

In nowadays, although the methodology of recognizing the world was divided into "holism" and "reductionism" sometimes, an appropriate understanding to the reality of complex thing could be an organic combination and synergism of both "holism" and "reductionism" instead of "confrontation" or "separation" of them. A sole "holism" will be empty without "reductionism" due to the absence of details inside, and an insular "reductionism" without "holism" will only see the movement of a single element but no information of the whole especially the attached—behavior due to interaction among "individual parts", like "seeing trees only but not forest". In fact, we always expect to see both trees and forest in a sense in viewpoint of systems science, including even effect of environmental / surrounding factors, so as to behave properly.

In order to achieve the main purpose of revealing cognition of the universality of systems theory thinking by using some ordinary physical phenomena and the analysis, following contents containing collective actions are involved, such as classical vibration of many bodies and wave, the extreme value problems in natural world, status of electrons in molecule and metals, Ising model in phase transition, elementary excitation in solid, probabilistic multi—objective optimization in a system and its applications including robust design and reliability design, effective medium approximation, certainty in uncertain phenomena, and some typical systemic approaches in broad fields. These all reflect cooperative / synergetic effects, the wholeness of group actions, and the collective phenomena in a system completely. The relevant methods and consequences of above issues could be used to deal with analogous problems in systems science and engineering fields.

The following contents are contained in this book:

Chapter 1 briefly describes the fundamental thinking and requirements of systems theory;

Chapter 2 pithily illuminates the thinking of systems science in classical vibration and wave, and two-body problems;

Chapter 3 concisely demonstrates extreme value problems in natural world as a whole in temporal or spatial scopes;

Chapter 4 presents statuses of electron in hydrogen and electrons in molecule and metals comparatively;

Chapter 5 explains Ising model in phase transition and elementary excitation in solid;

Chapter 6 briefly describes probabilistic multi-objective optimization in a system and its applications including robust design and reliability design;

Chapter 7 gives concisely description of effective media approximation in spirit of ergodic hypothesis in statistical physics;

Chapter 8 demonstrates certainty in uncertain phenomena;

Chapter 9 illuminates typical systemic approaches in broad fields;

Chapter 10 gives a comprehensive summary of the whole book.

Systems theory (systems science) is not a castle in the air but true to life, it originates from the life, so it must serve the life and engineering practices. There are idioms in China that "stones from other hills may be good for polishing this jade" and "comprehend by analogy". The living soul of Marxism is "seeking truth from facts", "abandoning dogma" and "concrete analysis". One should proceed from reality, constantly raise awareness, explore and reveal laws, so as to approach to the truth and to behave appropriately. The authors wish this work could play its role to make its contribution to public, research and specifically practical application of systems theory, so that it promotes the transformation of wisdom to productivity and makes our engineering practical work closer to success.

Professors Laike Tian and Kaiping Liu are acknowledged for their valuable discussions and suggestions. Thanks are given to Dr. Mengchu Huang of Springer for his helps in improving the contents.

This book can be used as a textbook for postgraduate and advanced undergraduate students in relevant majors, and a reference book for practicers in related fields.

Xi'an, China

Maosheng Zheng
Jie Yu

Contents

1 Outline of Systems Theory 1
- 1.1 Introduction 1
- 1.2 Basic Knowledge of Systems Theory 6
- 1.3 Inspiration 9
- 1.4 Summary 10
- References 10

2 Classical Vibration and Wave, and Two-Body Problems 13
- 2.1 Introduction 13
- 2.2 Vibration Problems of Simple Vibrators, and Two-Body 13
- 2.3 Vibrations of a One Dimensional Atomic Chain 15
- 2.4 Travelling Waves and Standing Waves 17
- 2.5 Effect of Nano-size on Vibration of One Dimensional Atomic Chain 20
- 2.6 Nonlocal Continuum Mechanics 21
- 2.7 Two-Body in Potential Field Problem 23
- 2.8 Certainty and Chaos 24
- 2.9 Summary 25
- References 26

3 Extreme Value Problems 27
- 3.1 Introduction 27
- 3.2 Extreme Value Problems in Electricity 28
- 3.3 The Regular Hexagonal Structure of the Hive 31
- 3.4 Energy Method 31
- 3.5 Least Action Principle 34
- 3.6 Applications of Energy Method in Solving Engineering Problems 38
- 3.7 Discussion on Overall Approach and Its Applications 45
- 3.8 Summary 46
- References 46

4 Characteristics of Electron in Atom, Molecule and Metal 49
4.1 Introduction 49
4.2 Electrons in Hydrogen and Helium Atoms 49
4.3 Electronic Gas in Metals, First Ionization Energy of Atoms and Work Function of Bulk Metals 52
4.4 Electrical Resistivity of Pure Metal 53
4.5 Summary 55
References 55

5 Ising Model in Phase Transition and Elementary Excitation 57
5.1 Introduction 57
5.2 Ising Model in Phase Transition 58
5.3 Elementary Excitation in Solid 61
5.4 Site-Bond Percolation and Phase Transition in Diluted Magnetic System 62
5.5 Order–Disorder Transition of Binary Ordered Alloys 66
5.6 Landau's Phase Transition Theory 68
5.7 Other Approaches Beyond Our Descriptions 70
5.8 Van der Waals' State Equation of Gas 72
5.9 Scale Transformation and Renormalization Group 72
5.10 Concluding Remarks 75
References 75

6 Brief Description of Probabilistic Multi-objective Optimization of a System 77
6.1 Introduction 77
6.2 Probabilistic Multi-objective Optimization 78
6.3 Idea of Preferable Probability and Its Quantization 79
6.4 Outline of Probabilistic Multi-objective Optimization 80
6.5 Simple Application Examples of PMOO 81
6.6 Robust Design as a Special Case of Probabilistic Multi-objective Optimization 89
6.7 Fuzzification as an Auxiliary Measure for PMOO Robust Design in Case of Desired Value as Target 93
6.8 Reliability Design Involving Optimization with Multiple Objectives by Means of Probabilistic Multi-objective Optimization 101
6.9 Discussion on Integral Approach and Its Applications 106
6.10 Concluding Remarks 108
References 108

7 Concise Description of Effective Media Approximation 111
7.1 Introduction 111
7.2 Effective Elasticity Modulus of Grains Reinforced Composite Materials 112
7.3 Strength and Toughness Parameters of Porous Materials 116
7.4 Effective Thermal Expansion Coefficient of Random Composite 118
7.5 Effective Thermal Conductivity, Dielectric Constant, Magnetic Permeability, and Electrical Conductivity of Composite with Random Phases 119
7.6 Analysis of Thermal Stress in Granular Composite 121
7.7 Summary 121
References 122

8 Certainty in Uncertain Phenomena 123
8.1 Introduction 123
8.2 Statistical Laws 123
8.3 Schrodinger's Cat and Its Enlightenment 129
8.4 Parallels Between Modern Physics and Eastern Mysticism 130
8.5 Goodhart's Law 133
8.6 Ranking of Fuzzy Numbers 134
8.7 Summary 136
References 136

9 Typical Systemic Approaches in Broad Fields 139
9.1 Introduction 139
9.2 Lotka-Volterra System of Ecology 140
9.3 "Five Essential Elements" System of Ancient China 141
9.4 Method of Electric Images in Electrostatics 143
9.5 Application of Nonlocal Euler–Bernoulli Beam Theory in Static Microstructure 149
9.6 Summary 152
References 153

10 General Conclusions 155

About the Authors

Dr. Maosheng Zheng received his Bachelor's degree in Theoretical Physics from Northwest University, Xi'an, China, in 1983, Master's degree in Electronic Physics from Xidian University, Xi'an, China, in 1985, and Ph.D. degree in Materials Science and Engineering from Northwestern Polytechnical University, Xi'an, China, in 1992. From June 1994 to June 2006, Dr. Zheng was a Professor at the School of Materials Science and Engineering, Xi'an Jiaotong University, and since June 2006, Dr. Zheng has been a Professor at the School of Chemical Engineering, Northwest University.

Dr. Zheng focuses on the research of *probabilistic multi-objective optimization (PMOO), robust design*, and *materials science and technology*. Till 2024, Dr. Zheng has published seven monographs (1st authored six, Springer four, Eliva Press one and Chemical Industry Press one; coauthored one, Science Press) more than 300 peer-reviewed papers in international journals and conferences and chaired "863" project and many projects of Science and Technology Ministry and Shaanxi Province of China. He is a Member of China Energy Society, Vice Chairman of the Shaanxi Corrosion and Protection Society, and served as a reviewer of several international journals. He was a recipient of State Council Special Allowance, 2nd prize of Science and Technology Progress Award of China Aviation Industry Corporation, 3rd prize of Science and Technology Progress Award of the Ministry of Education, 2nd prize of Huo Yingdong Young Teacher (Research), 2nd Youth Science and Technology Award of Shaanxi Province, and 2nd prize of the Natural Science Award of China Higher Education, respectively.

Jie Yu received her Bachelor's degree in Polymer Science and Technology from Northwestern Polytechnical University, Xi'an, China, in 1987. In July 1987 to December 1992, Yu worked at Xi'an Far East Machinery Manufacturing Company as an Assistant Engineer responsible for design and processing of rubber and plastic parts; in January 1993 to September 2007, Yu served as an Engineer in Department of Polymer Materials of Xi'an Jiaotong University; in September 2007 to November 2011, Yu worked as an Engineer at School of Life Sciences of Northwest University; in December 2011 to July 2020, Yu worked as a Senior Engineer at School of Life Sciences of Northwest University; since August 2020, Yu served as a Professor-leveled Senior Engineer at School of Life Sciences of Northwest University.

Jie Yu focuses on *design of experiment and optimization*, *research of polymer materials and their applications in medicine.* Till 2024, Yu has published more than 60 scientific research papers, coauthored five monographs (Springer three, Eliva Press one, and Chemical Industry Press one), chaired and completed two key projects of Shaanxi Province, participated in the completions of five national and provincial key projects, and two enterprise cooperation projects, obtained national invention patents three items, and two items have been transformed to serve the society, won one advanced worker of Northwest University, two Science and Technology Awards of Shaanxi Provincial Higher Education Institutions, and one gold medal of Shaanxi Science and Technology Workers Innovation and Entrepreneurship Competition.

Chapter 1
Outline of Systems Theory

Abstract Fundamentals of systems theory are outlined in this chapter, it aims to extract guidable and valuable ideas.

Keywords Systems theory · Wholeness · Component · Irreplaceability · Organic integral

1.1 Introduction

Fundamentals of systems theory are the indespensible issues for research and application of systemic thinking. So, in this chapter the relevant contents of systems theory are outlined. The problems in systems theory are commonly interdisciplinary with complexity. But models and abstractive extractions are usually useful and powerful tools [1].

The root of systems theory has a long history. The simple thought of systems theory sprouted in the Spring and Autumn Period of ancient China (770-476 BC) and Demokritos of ancient Greece (about 460-370 BC).

In ancient China, there was preliminary thinking of systems theory. The typical stories of "The Art of War" of China in the Spring and Autumn Period (515-512 BC) and "Tian Ji's Horse Racing" in the Warring States Period (357-354 BC) are considered to have the spirit of primitive systemic thinking. An overall optimization thought was involved in above systemic problems. Another famous example is the Dujiangyan Water Conservancy Project organized by Li Bing and his son, it was in the Warring States Period (256-251 BC), the oldest large-scale project was implemented in engineering, which has been playing a role till now. It is not only an important resource for agriculture in the plain of western Sichuan, but also protects local agriculture from the early waterlogging disaster, and arranges nearly 1000 small and medium-sized hydropower stations [2–4]. From the exquisite site selection of Dujiangyan Water Conservancy to the overall coordination of various facilities, key problems such as automatic water regulation, sediment discharge and flood discharge

M. Zheng and J. Yu, *Systems Theory for Engineering Practice*,
https://doi.org/10.1007/978-981-97-9342-6_1

have been successfully solved, and a series of "systems engineering" with high reliability, low cost, great benefit and simple maintenance have been formed to achieve system optimization as a whole. It embodies the idea of "systems engineering" that integrates management, construction and maintenance.

Democritus, a materialist philosopher in ancient Greece, once put forward the concept of "cosmic large system" and used the word "system" at the earliest. Heraclitus, one of the founders of dialectics, believes that "the world is a whole that includes everything"; Later generations attributed Aristotle's famous words to "the whole is greater than the sum of its parts", which is one of the basic principles of systems theory.

From their long-term production and life, the sages of ancient China abstracted "five essential elements" (or called Wuxing), namely, fire, earth, gold, water and wood to describe the "law" of complex movement and evolution of things in an organic body. Besides, the fundamental relationships of mutual "production" and "restriction" exist among the "five essential elements", such as: *fire generates earth, earth generates gold, gold generates water, water generates wood*, and *wood generates fire*; in contrary, *water restrictes fire, fire restrictes gold, gold restrictes wood, wood restrictes earth*, and *earth restrictes water*. It is undoudt that the five abstractive "elements" could be designated as A, B, C, D and E as well provided that the fundamental correlations of mutual "generation" and "restriction" relationships among them are met. This is a simple systemic view in ancient China. The "generation" or "production" here means benefit, including parenting, nourishing, increasing, promotion, etc.; the "restriction" indicates weakening and resistance. Mutual generation and mutual restriction are two opposing aspects in the system, so as to ensure the dynamic equilibrium state in the system synegetically. The relationship between any two "elements" is not a simple "one-way relationship", but mutual. Any one of the five elements has four relationships: giving birth to me, my giving birth to it, resisting me and my resisting it, so as to guarantee the normal operation of the "restraint" relationship. A stronger "fire" could restrict or weaken "gold" excessively, while the weaker "gold" could not restrict or weaken "wood" sufficientely, thus "wood" grows strongly, and the stronger "wood" could restrict or weaken "earth" excessively; the weaker "earth" could not restrict or weaken "water" effectively, as a result, there appears excessive "water". Furthermore, the excessive "water" could supply plenitudinous restriction or resistance to the stronger "fire", so as to make the system return to its normal state. This is the rule. The affected "element", through some intermediate links, acts on the actor in turn, that is, it produces feedback adjustment action (effect), thus ensuring the relative balance of the system. The theory of "five essential elements" uses the relationship among the five elements to clarify the relationship inside things, and it holds that nothing is isolated or static, but is in a constant movement of mutual generation and mutual restriction to maintain its dynamic balance. This theory has gained successful applications in traditional Chinese medicine and health care widely.

Figure 1.1 represents the diagram of correlations among the "five esential elements". The diagram is meaningful, which has *twofold meanings* at least, i.e.,

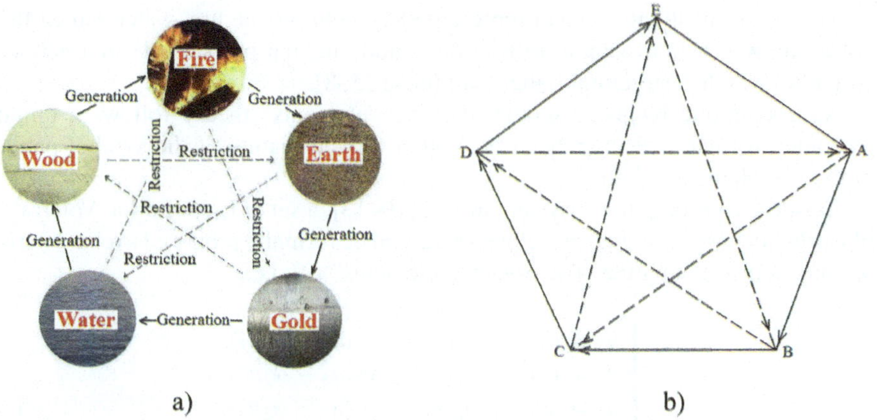

Fig. 1.1 Mutual correlations of the "five essential elements" in ancient China

the specific number "five", and the fundamental correlations of mutual "generation" and "restriction" among the "five esential elements" [5–8].

As to the specific number "five", it was proved in recent years that the specific number "five" is an unique number, which is with completeness under conditions of mutual "generation" and "restriction" among the "five esential elements" from the perspectives of topology theory, graph theory and group theory [5–8]. In addition, the specific number "five" is a prime number consisting of the sum of the first even number and odd number, which are the top two prime numbers. Besides, the characteristics of the number "five" are considered as "harmony of yin (negative) and yang (positive), success and vitality" due to its nature of containing the first even number (yin) and the first odd number (yang). The regular pentagon and pentagram connected by five points include the golden section ratio, which has infinite wisdom and vitality.

The other half of the twofold is that the fundamental correlations of mutual "generation" and "restriction" among the "five esential elements", which could be miraculously revealed by the edges of the five acute triangles and five obtuse triangles in Fig. 1.1b, i.e., *ACD*, *BDE*, *CEA*, *DAB* and *EBC*, as well as *ABC*, *BCD*, *CDE*, *DEA* and *EAB*.

The edges of each triangle represent a specific meaning of mutual correlations. The *solid line of the edge reflects the "generation" along the arrow, the dashed line reveals the "restriction" along the arrow* in Fig. 1.1b.

Such triangle relationship is a complex relationships based on simple correlations, which can be called a restrict relationship. They still abide by the fundamental principles of the five elements. By introducing another item inside the system, the restriction is complicated, and the one-way relationship of machinery becomes a dialectical relationship affected by quantity and specific conditions.

That is to say, the movement represented by gold, wood, fire, water and earth is not a one-way action without restriction conditions, but plays a role in a network structure, which is structural rather than linear [5–8].

As a symbolic language model, the five "elements" theory follows a unique reasoning format, and illogicality is another typical feature of the symbol pattern of the five elements [5–9].

Besides, according to ecosystem model [10–13], a set of L–V (Lotka–Volerra)—like primary system of dynamic equations could be formally proposed to characterize the interaction among the "five essential elements" [9], i.e.,

$$\begin{cases} \dot{u}_1 = u_1(\varepsilon_1 + \eta_{15}u_5 - \gamma_{14}u_4 - \alpha_1 u_1), \\ \dot{u}_2 = u_2(\varepsilon_2 + \eta_{21}u_1 - \gamma_{25}u_5 - \alpha_2 u_2), \\ \dot{u}_3 = u_3(\varepsilon_3 + \eta_{32}u_2 - \gamma_{31}u_1 - \alpha_3 u_3), \\ \dot{u}_4 = u_4(\varepsilon_4 + \eta_{43}u_3 - \gamma_{42}u_2 - \alpha_4 u_4), \\ \dot{u}_5 = u_5(\varepsilon_5 + \eta_{54}u_4 - \gamma_{53}u_3 - \alpha_5 u_5). \end{cases} \tag{1.1}$$

In Eq. (1.1), u_i ($i = 1, 2, \ldots, 5$) reflects the representative quantity of i-th essential element; ε_i ($i = 1, 2, \ldots, 5$) indicates the self incremental factor of i-th essential element; η_{ij} ($i = 1, 2, \ldots, 5; j = [(i + 4) \bmod 5]$ expresses the representative generation factor of j-th essential element to i-th essential element, in which $[(i + 4) \bmod 5]$ expresses the integer remainder of number $(i + 4)$ divided by 5; γ_{ij} ($i = 1, 2, \ldots, 5; j = [(i + 3) \bmod 5]$ indicates the representative restraint factor of j-th essential element to i-th essential element; α_i ($i = 1, 2, \ldots, 5$) indicates the representative repelling factor of i-th essential element in itself mutually. Till now we have not known the exactly quantitative values of above representative factors. There was also a similar formula including unknown terms of "injection signals and fluctuations" [14].

Especially, a case of the system is at its stable status, thus the dynamic equations for the "five essential elements" is like this, $\dot{u}_1 = \dot{u}_2 = \dot{u}_3 = \dot{u}_4 = \dot{u}_5 = 0$. In this specific condition, it derives,

$$\begin{cases} \varepsilon_1 + \eta_{15}u_5 - \gamma_{14}u_4 - \alpha_1 u_1 = 0, \\ \varepsilon_2 + \eta_{21}u_1 - \gamma_{25}u_5 - \alpha_2 u_2 = 0, \\ \varepsilon_3 + \eta_{32}u_2 - \gamma_{31}u_1 - \alpha_3 u_3 = 0, \\ \varepsilon_4 + \eta_{43}u_3 - \gamma_{42}u_2 - \alpha_4 u_4 = 0, \\ \varepsilon_5 + \eta_{54}u_4 - \gamma_{53}u_3 - \alpha_5 u_5 = 0. \end{cases} \tag{1.2}$$

The solution of Eq. (1.2) expresses the synergetic and coordinate correlations among the "five essential elements" in stable condition.

Friedrich Engels once said that when many people cooperated, many forces would merge into a general force, and in Karl Heinrich Marx's words, it would create a 'novel force', which will differ from the simple summation of its essential forces.

In the modern sense, systems science appeared its embryonic form in the 1940s, and came out in the 1960s. By the 1970s, not only the components were further perfected, but also a preliminary structural framework was built, and systems science was built as a knowledge system as a whole [15]. While, Capra et al. pointed out that twenty to thirty years before Bertalanffy's publication of his papers on "general systems theory" [16], a Russian specialist, Alexander Bogdanov, proposed a systems theory of equal sophistication and scope, which has been still unknown outside of Russia.

The remarkable characteristic of systemic thinking is the wholeness and advocate understanding things in solving problems as a whole. When applying systems science to study and solve actual problems, it emphasizes that we should not stop at understanding various disciplines, but conduct the things as a complete system that composed of these branches and master it as a whole. The task of systems science is to provide an overall methodology and explanation to things.

Through decades of research and application, systemic thinkings and methods have been integrated into the knowledge structure of workers in natural science, engineering technology, social science, business management and other fields. From public to specialists, from academic journals to literary works, terms such as system, information, systems engineering and self-organization are frequently used. Although there are still differences, the status of systems science as a discipline has been basically recognized by more and more people to consider and solve problems.

In recent decades, more attention has been paid to using mathematical tools and mathematical models to depict and give precise solutions to systemic problems. This is the way for systems science to mature. But at the same time, it is not simply pursuing the profound beauty of mathematical tools, ignoring the actual background and empirical significance of the problem, and weakening the systemic significance. On the contrary, it is necessary to emphasize its systemic significance.

Besides, the research of systems science theory itself needs to be deeply explored. It is necessary to analyze new phenomena academically and problems in application. Therefore, it is necessary to understand fundamental knowledge of systems theory.

In recent years, Capra et al. emphased that evolution cannot be seen as a competitive struggle for existence, but rather as a cooperative dance in which creativity and the constant emergence of novelty are the driving forces [16]. This is a viewpoint of harmonious coexistence of things within a system obviously, therefore maybe the dynamic equations for the "five esential elements" of ancient Chinese sages is beneficial and useful in a sense.

In the world, things are usually complex, or even with uncertainty in viewpoint of systems science. Probability theory is a mathematical tool to study the regularity of uncertain phenomena, which came into being in the middle of seventeenth century. It has not only its own independent research problems, but also extensive applications in the real world. Since the twentieth century, probability theory has gradually penetrated into almost ubiquitous fields such as natural science, social science and people's daily life. It plays an increasingly important role in both research and application fields.

The development of probability theory underwent different stages. The classical probability theory, combinatorial probability theory, was born in the middle of seventeenth century and formed in 1812, algebraic methods are mainly used. Modern probability theory, analytical probability theory, was formed from 1812 to the beginning of the twentieth century, analytical methods are mainly based on. Contemporay probability theory, measurement probability theory, was formed after 1933, it mainly studies probability theory with measurement theory, which can be called measurement probability theory. On the basis of axiomatization, contemporay probability theory has made a series of theoretical breakthroughs and rapid development [17].

1.2 Basic Knowledge of Systems Theory

Systems science and systems engineering are the inevitable products of human social production practice and the development of science and technology. Systemic thought is an important component of dialectical materialism as well [18–20]. The development of modern science and technology has a great influence on systemic thinking and method. Systemic thought is a dialectical tool for analysis and synthesis, and through quantitative description and continuing revision/improvement in the practical application of systems engineering, a "methodology" could be formed, in which systems thought could be transitted into a specialized science-system science gradually.

Ludving von Bertalanffy, an Austrian-American theoretical biologist, considered that a system is a "synthesis of interacting elements". The general definition of system is that a system is an organic whole with specific functions, which is composed of several components (elements) that is with interaction and interdependence.

It should meet three conditions in a system: (1) The system must be composed of two or more components (parts and elements), and each part is the most basic unit of the system, which can also be regarded as the carrier and foundation of the system; (2) There is a certain organical connection between each parts surely, which makes the exterior and interior of the system form a certain order or structure; (3) Every system must have a specific new function beyond that of its constituent elements, which is determined by the organic connection and composition of all parts of the system.

The unity of opposites within the system and the parts, as well as within the components, enables the system to play its special functions.

Systems can be classified into different types according to their forms generally, such as,

(1) Natural systems and man-made systems;
(2) Entity system and conceptual system.

A physical system is that composed of various physical substances, and its concrete manifestation is mainly hardware. While a conceptual system is that composed of concepts, principles, laws, methods, rules, procedures and other conceptual immaterials, which is mainly expressed in the form of "software", such as elastic modulus, elongation, program system, specification system, etc.

(3) Closed system and open system;

(4) Static system and dynamic system;
(5) Object system and behavior system;
(6) Control system and causal system.

Main characteristics of the system

The key to understand, study and master "systemic thought" is to get at the features of the system. In general, the system was considered to have four main characteristics: integrity, relevance, purpose and environmental adaptability.

1. Integrity (wholeness)

The integrity of a system is mainly displayed in the overall function of the system. The overall function of the system is not a simple superposition of the roles of each component, nor is the simply pile up of the components. The overall function of a system refers to a novel function. This feature is commonly summarized as "the whole system being not equal to the simple sum of its components", but "the whole being bigger than the sum of its parts".

The rational deductions of integrity of a system are: (a) There is irreplaceability of each part which appears in the system simultaneously; (b) The optimization of the system is the optimization of the whole system, which could not be set as the accumulation of the optimization of its various components.

Each component is an organic part of the system, which has its goal and function. The optimization of a system is to make the system in the best state as a whole, all its parts/components could be "united" under the banner of "overall optimization" and "working" in coordination. Only when all components in the system are coordinated with each other, can the comprehensive optimization of the system objectives and the optimization of the functions of the whole system be realized.

2. Correlation

Each part of the system is interactive and interrelated. Relevance is to show the organic correlations, connections and constraints between these components. If a certain part of the system changes, other related parts will give a response or adjustment inevitably, so as to ensure the system in the overall working state.

3. Purpose

The purpose of the system refers to the goal decided by the needs of its nature and people's practice.

Sometimes the purpose of artificial systems is not a sole one.

4. Environmental adaptability

Environmental adaptability refers to that the system needs to respond or adjust to the changes of "things outside" (information, matter and energy) in order to adapt to the changes of the environment.

In short, from viewpoint of systems theory, we can make a comprehensive analysis and deployment of the system.

Under the guidance of systems theory, systems engineering is to realize the effective "operation" of the system with the integration of research ideas, the integration of application methods, and the scientific and modern organization and management. These are also its essence and characteristics. Certain factors are its internal causes in one system, but they might be divided into external causes under the division of another system [21]. If the factors which determine the evolution of the system are divided outside the system, they can be regarded as "external control" and can be analyzed by using "outside organization theory". When the factors which control the evolution of the system are divided into the system, we can use the "self-organization theory" to analyze the relationship between the internal factors.

In general, the reasons which affect the evolution of the system may be regarded as either external factors or internal factors according to the ways of dividing the system, it could result in the choice of external organization theory or self-organization theory for further research and analysis. The way of division should be selected properly to ensure the ease of actual research and analysis [21].

Figure 1.2 shows an example of a system G including components $g_1, g_2, \ldots, g_5$, and their correlations r_{ij} between components g_i and g_j in general.

Self-organization theory was proposed by Synergetics founder, German scientist Haken [22], the word "synergy" refered to the individual actions which are beyond

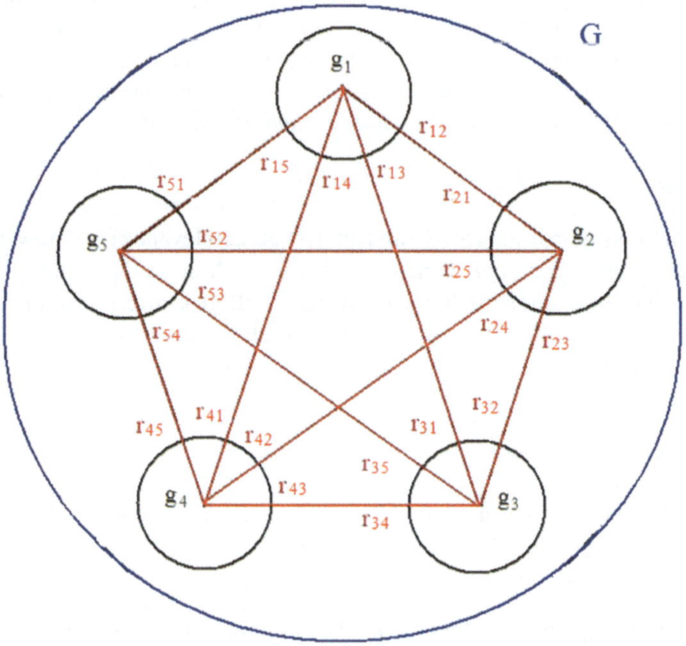

Fig. 1.2 System G including components $g_1, g_2, \ldots, g_5$, and their correlations r_{ij}

the parts of a system, thus constituting the unified function of the whole system. Synergy is an inherent self-organizing function of any system itself.

The optimization of the system is the overall optimization of the system, which does not equal to the accumulation or pile-up of the optimization of its components [9, 23].

1.3 Inspiration

As to model, we remember Box's famous statement that "all models are wrong, but some are useful" [24–28]. Modeling can play a role of analogy. This can be seen as a convenient routine to justify the effectiveness of a modeling. A key insight is to take the modeling process and the relevant predictions as a "way of thinking" to recognize the complex system in some respects. One model is a model, which is only an expression of understanding reality from analogical or abstractive measurements in some respects; of course it is not the object entity or even an actual representation of the reality. On the other hand, one's understanding and expression in words are always limited than the real complexity of actual thing, so it could not expect any model to fully reveal the reality of whole thing in this point. "A blind feeling an elephant" is a very common thing, only some partial apparent characteristics could be felt instead of reality itself, but human's understanding and recognition could be deepened gradually. Actually wisdom is productivity, which highly abstracts the essence of things and gains insight into the development and evolution of things. Box's phrase emphasizes the value of abstraction in order to gain insight, while the abstraction is valid within a certain range clearly in incomplete description, which is enough for our usage in certain range intrinsically. This provides a way of thinking about complex systems and mechanisms underlying observations and practices.

In epiestemology, there appeared two methodologies, i.e., the so-called "holism" and "reductionism" sometimes. Actually, an appropriate understanding to the reality of complex thing could be an organic combination and synergism of both "holism" and "reductionism" instead of "confrontation" or "separation" of them. A sole "holism" will be empty without "reductionism" due to the absence of details inside, and an insular "reductionism" without "holism" will only see the movement of a single element but no information of the whole especially the attached-behavior due to interaction among individual constituent elements, like "seeing trees only but not forest". While it is expected to see both trees and forest including even effects of surrounding/environmental factors in a sense in viewpoint of systems science so as to make our engineering practices closer to success.

Besides, it is quite important to view and apply probability theory rationally for solving practical problems. According to Jaynes [29], every probability assignment is 'subjective' appropriately, which only reflects a state of knowledge instead of actual measurement in a physical experiment. The knowledge is based on the database to be used to perform the derivations.

But in an actual evaluation it was just the function of our interface desiderata to make the probability assignment completely 'objective' in the sense, which are independence of user's personality. It is a means of reflecting the *information* given in the statement of a problem, which is independent on whatever personal feelings, hopes, fears, or value judgments, etc. So, it is 'objectivity' in this sense that needs a respectable theory of inference scientifically.

More than 100 years ago, in his youth, Chairman Mao Zedong, the great leader of the people of China, wrote a glorious poem, "confident in life for two hundred years, will be hitting water for three thousand miles" to express his willing to fight bravely despite difficulties. Li Bai, a famous poet in China in the Tang Dynasty, wrote a heroic poem "I am born with talent and must be useful, a thousand pieces of gold will be scattered and come back again", which also reflects his life values. We believe that these spirits will inspire countless scientific and technical workers to make unremitting efforts and exploration in philosophic and practical problems.

1.4 Summary

From all above concise studies on systems theory, it extracts following valuable substances.

(1) The remarkable feature of systemic thought is that it emphasizes wholeness of system, and each part that appears simultaneously in the system is irreplaceabile;
(2) The optimization of the system is the overall optimization of the whole system, which is the synergetic actions among its consisting components and against the any type of simple accumulation or pile-up of the optimization of its components.
(3) The involved methodologies for systems theory are organic combination and synergism of both "holism" and "reductionism" instead of "confrontation" or "separation" of them, which could be used in dealing with analogous problems in systems science-engineering fields.
(4) The aim of this work is to play its role as a stone from other hills but servicing to polish this jade, which makes its contribution to public, research and specific application of systems theory so as to promote transformation of wisdom to productivity and make our engineering practical work closer to success.

References

1. J.S. Wu, *Invitation to Systems Science* (Science Press, Beijing, 2018)
2. Z. Deng, L. Fang, A model of applying systems engineering thought in ancient China—Dujiangyan water conservancy project. Theory Pract. Syst. Eng. **7**(1), 57–61 (1987)
3. F. Wang, H. Li, J. Ding, Y. Chen, Systems dialeetieal principles in Dujiangyan irrigation works. J. Syst. Dialectics **7**(1), 55–60 (1999)

4. G. Chen, A Summary about studies of Dujiangyan water conservancy since 1970s. J. Guangxi Univ. Nationalities (Natural Sci. Ed.) **22**(2), 39–43 (2016)
5. H. Qian, The number of five in "Wuxing: a new interpretation. J. Beijing Normal Univ. (Soc. Sci. Ed.) **52**(4), 72–78 (2007)
6. L. Wang, W. Wei, Preliminary study on topological principles of five elements theory. J. Tradit. Chin. Med. Lit. **18**(3), 23–24 (2000)
7. H. Hua, The theory of five elements—the symbolic system in ancient China. J. Dialectics Nature **17**(3), 48–57 (1995)
8. X. Wang, S. Liu, Wuxing system and group structure. Math. Pract. Theory **39**(18), 177–182 (2009)
9. M. Zheng, J. Yu, *Robust Design and Assessment of Product and Production by Means of Probabilistic Multi-objective Optimization* (Springer, Singapore, 2024)
10. H. Marcus-Roberts, M. Thompson, *Life Science Models* (Springer, New York, 1983)
11. J. Muller, C. Kuttler, *Methods and Models in Mathematical Biology, Deterministic and Stochastic Approaches* (Springer, Germany, 2015)
12. W. Lin, *Ecology*, 2nd edn. (Science Press, Beijing, China, 2024)
13. Y. Lin, R. Fu, *Ecology*, 3rd edn. (Science Press, Beijing, China, 2023)
14. F. Li, Functional structure model of human body and Yinyang-Wuxing equations. Phys. Scr. **36**, 966–969 (1987)
15. D. Miao, *Essentials of System Science*, 4th edn. (Renmin University of China Press, Beijing, 2016). ISBN: 978-7-300-23289-8
16. F. Capra, P.L. Luisi, *The Systems View of Life: A Unifying Vision* (Cambridge University Press, Cambridge, UK, 2014)
17. C. Zhu, *Historical Research on the Thought and Method of Probability Theory* (University of Electronic Science and Technology of China Press, Chengdu, 2007). ISBN: 978-7-81114-619-6
18. L. Yang, *Systems Engineering Methods and Applications* (Metallurgical Industry Press, Beijing, 2018)
19. H. Li, C. Zhang, *Systems Engineering and Application* (Machinery Industry Press, Beijing, 2013)
20. J. Wu, *Methodology and Application of Modern Systems Science* (University of Electronic Science and Technology Press, Chengdu, 2005)
21. G. Xu, *Systems Science* (Shanghai Science and Technology Education Press, Shanghai, 2000)
22. H. Haken, *Advanced Synergetics, Instability Hierarchies of Self-organizing System and Devices* (Springer, Berlin, 1983)
23. M. Zheng, J. Yu, H. Teng, Y. Cui, Y. Wang, *Probability-based Multi-objective Optimization for Materials Selection*, 2nd ed. (Springer, Singapore, 2023). https://doi.org/10.1007/978-981-99-3939-8
24. G.E.P. Box, N.R. Draper, *Empirical Model-building and Response Surfaces* (Wiley, New York, 1987)
25. G.E.P. Box, Robustness in the strategy of scientific model building, in *Robustness in Statistics*, ed. by R.L. Launer, G.N. Wilkinson (Academic Press, New York, 1979), pp. 201–236
26. G.E.P. Box, Science and statistics. J. Am. Stat. Assoc. **71**(356), 791–799 (1976)
27. H. Enderling, O. Wolkenhauer, Are all models wrong? Comput. Syst. Oncol. **1**, e1008 (2021). https://doi.org/10.1002/cso2.1008
28. L. Wen, D. Tuffley, T. Rout, Using composition trees to model and compare software process, in *Software Process Improvement and Capability Determination 11th International Conference, SPICE 2011, CCIS 155*, ed. by R.V. O'Connor, T. Rout, F. McCaffery, A. Dorling (Springer, Berlin, Heidelberg, 2011), pp. 1–15
29. E.T. Jaynes, *Probability Theory the Logic of Science* (Cambridge University Press, Cambridge, 2003)

Chapter 2
Classical Vibration and Wave, and Two-Body Problems

Abstract The typical problems in classical vibration and wave, i.e., vibrations of two bodies connected with a spring, vibrations of an atomic chain, mechanical waves on a line, and effect of nano-size on vibration of one dimensional atomic chain, nonlocal continuum mechanics, etc., as well as two-body and three-body problems, exhibit group behavior of a system in whole, proper models are needed to characterize them.

Keywords Classical vibration · Wave · Atomic chain model · Nonlocal continuum mechanics · Group behavior · Collective phenomenon · System

2.1 Introduction

In classical physics, there are many phenomena which exhibit group behavior of a system actually, such as, vibrations of two bodies connected by a spring, vibrations of an atomic chain, waves, etc., as well as nonlocal continuum mechanics. Appropriate models are introduced to describe these phenomena, which forms classical vibration and wave theories of behavior of the system in a whole, and can be useful to demonstrate analogous case in systems science as proper approach [1–6]. While, the "coupled" movements of the relevant bodies are presented in the two-body and three-body problems in the system distinctly in classical physics.

2.2 Vibration Problems of Simple Vibrators, and Two-Body

1. A vibrator connected with a spring

To begin with, a vibrator connected with a spring is shown in Fig. 2.1 as a prelude example. The mass of the body is m, the spring is with a stiffness coefficient k fixed at point O. Press the spring together with the body by a small distance, and then release them, a harmonic vibration is thus initiated. The vibration can be described by Eq. (2.1) [1–6],

M. Zheng and J. Yu, *Systems Theory for Engineering Practice*,
https://doi.org/10.1007/978-981-97-9342-6_2

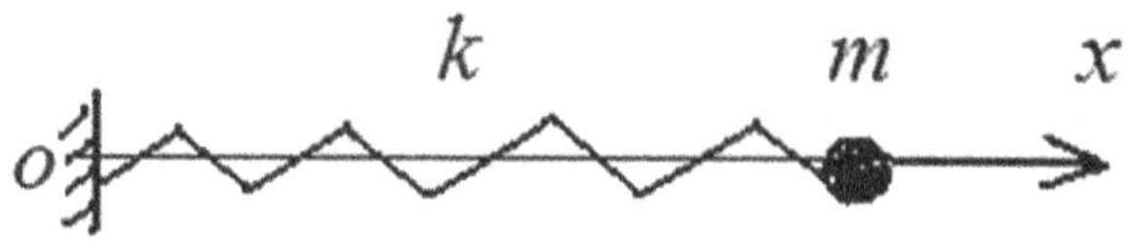

Fig. 2.1 A vibrator connected with a spring

$$x(t) = A\cos(\omega_0 t + \varphi_0), \omega_0 = \sqrt{k/m}. \tag{2.1}$$

In Eq. (2.1), $x(t)$ is the displacement of body at instant time t with respect to its position in equilibrium status; A is the amplitude; φ_0 is the initial phase; ω_0 is its circular frequency, $\omega_0 \equiv 2\pi f \equiv 2\pi/T = \sqrt{k/m}$, f_0 is its frequency, T is period of the vibration. The resilience due to Hooke's law is $F = -kx$. Obviously, the harmonic vibration is with the circular frequency ω_0 of $\omega_0 = \sqrt{k/m}$.

2. **A vibrator of two bodies connected with a spring and its application to vibrations of a diatomic molecules**

A vibrator of two bodies, m_1 and m_2, is connected at their ends by a spring with a stiffness coefficient k, which is shown in Fig. 2.2. Press the spring together with the bodies by a small distance, and then release them. Rather complex movements including harmonic vibration are thus established. The vibrations can be described by Eqs. (2.2) and (2.3) [1–6],

$$x_1(t) = A_1\cos(\omega_a t), \tag{2.2}$$

$$x_2(t) = A_2\cos(\omega_a t + \pi). \tag{2.3}$$

In Eq. (2.2), $x_1(t)$ is the displacement of body 1 at instant time t with respect to its position in equilibrium status; $x_2(t)$ is the displacement of body 2 at instant time t with respect to its position in equilibrium status; A_1 and A_2 express the amplitudes of body 1 and body 2, respectively; ω_a is its circular frequency, $\omega_a = \sqrt{k/\mu}$, in which $\mu(\mu \equiv m_1 m_2/(m_1 + m_2))$ expresses a reduced mass of bodies 1 and 2; $A_1/A_2 = -m_2/m_1$, expresses the vibrations of both bodies are in opposite directions.

Another solution is with the circular frequency $\omega_b = 0$, which indicates a rigid movement of the two bodies as a whole [1–6].

Obviously, the harmonic vibration is with the new circular frequency ω_a, i.e., $\omega_a = \sqrt{k/\mu}$, which differs from that of the vibrator of single body distinctly, see Eq. (2.1), and the influences of interaction between two bodies are included by the

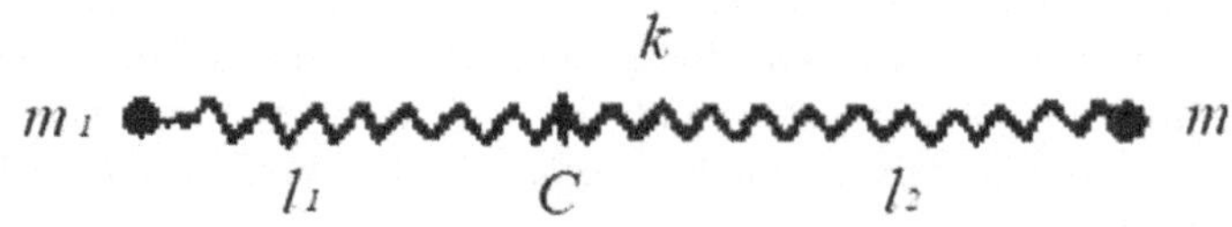

Fig. 2.2 A vibrator of two bodies connected with a spring

Table 2.1 Some data for k, μ and f for a few chemical bonds

Chemical bond	Reduced mass μ/amu	Molecule	k/ × 10^5 dyn/cm	f/ × 10^{10} Hz
H–Cl	0.972	HCl	5.15	8664
H–O	0.941	H_2O	7.80	11,250
H–C	0.923	HH_3X	4.7	8820–9120
C–O	0.857		5.0–5.8	3336–3594
C–C	6.0		4.5–5.6	3594
C=C	6.0	C_2H_4	9.5–9.9	8043
C=O	6.857		11.8–13.4	5127–5463

new parameter μ—reduced mass of bodies 1 and 2. Besides, the rigid movement of the two bodies as a whole is not appeared in the problem of the vibrator of single body evidently.

As an application of vibrator of two bodies connected with a spring, the problem of vibrations of a molecule with two atoms, such as H_2, O_2 or N_2, H–Cl, C–C, C=C, H–O, C–H, is studied. Table 2.1 shows some data for k, μ and f for a few chemical bonds.

Vančik emphasized that water molecule has particular properties which are not the properties of consisting atoms, hydrogen and oxygen. As a result, properties of water cannot be predicted solely with the knowledge of the natures of both oxygen and hydrogen atoms. *These new unpredictable properties of water emerge only until the water molecule, i.e. the structure on the higher complexity level, is formed* [7]. *Such property of the whole shows more than any simple superposition of properties of the consisting components, which reflects the so called holism.* Actually, above systemic phenomena and approaches in holism appeared in physics are much earlier than Bertalanffy's papers on "general systems theory".

Capra discussed the difficulty of division of sub-atom particle into its "consisting particles" in the meanings of classical physics [8], which indicates the importance of overall cognition in sub-atomic field instead of division of sub-atom particle into its consisting particles individually.

2.3 Vibrations of a One Dimensional Atomic Chain

The vibrations of an atomic chain are presented here in this section.

The vibration of real lattice is a very complicated problem. However, the vibration of one-dimensional atomic lattice (such as one-dimensional monatomic lattice and one-dimensional diatomic lattice) is not only simple and solvable, but also reflects the basic characteristics of lattice vibration, and relevant methods and main consequences can be extended to three dimensions. In order to simplify the problem further, it assumes that only the interaction of nearest neighbor atoms is considered, that is, the so-called *nearest neighbor approximation*.

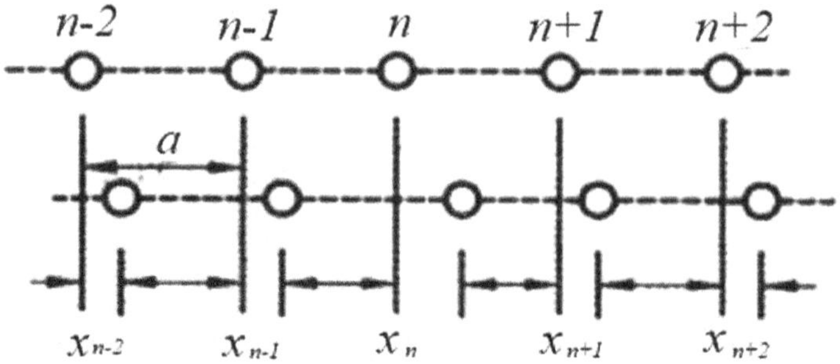

Fig. 2.3 A one-dimensional atomic chain

It further considers a one-dimensional atomic chain, as is shown in Fig. 2.3. Each atom has the same mass m, and the atomic spacing is a (lattice constant) at the equilibrium position [9, 10]. Due to thermal motion, each atom leaves its equilibrium position. Let x_n represent the displacement of the n-th atom from the equilibrium position. Thus, the relative displacement between the n-th atom and the $n+1$-th atom is $x_{n+1} - x_n$, and the motion equation of the n-th atom can be written as [9, 10],

$$m\frac{d^2x_n}{dt^2} = \beta(x_{n+1} - x_n) - \beta(x_n - x_{n-1}) = \beta(x_{n+1} + x_{n-1} - 2x_n). \tag{2.4}$$

In Eq. (2.4), β is with a resilience force coefficient, $\beta = d^2U(x)/dx^2|_{x=a}$, $U(x)$ is the interactive potential between atoms; m represents mass of atom. The resilience due to Hooke's law is $F = -\beta \cdot \Delta x$.

An equation can be written for each atom, so, a set of differential equation group consisting of n equations can be thus obtained. The number of equations is equal to the number of atoms N. We notice that there is a correlation between these atoms, and the frequency of atomic vibration is no longer equal to that without mutual coupling ($\omega_0 = \sqrt{\beta/m}$), but the solution of the equation can still be written as a plane wave with wave vector $\boldsymbol{q}$ and frequency ω, which is called "lattice wave", that is [9, 10],

$$x_n = Ae^{i(qna-\omega t)}. \tag{2.5}$$

When it satisfies Eq. (2.5), it is required that the frequency and wave vector of atomic vibration meet a certain relationship, that is, the dispersion relationship. Substituting expression of x_n into Eq. (2.4), we can get the dispersion relation as follows,

$$\omega = 2\sqrt{\frac{\beta}{m}} \cdot \left|\sin\frac{qa}{2}\right|, q = \frac{l}{N} \cdot \frac{2\pi}{a}, l = 0, \pm 1, \pm 2, ..., \pm N/2, \tag{2.6}$$

The lattice wave in the crystal given by Eq. (2.5) is shown in Fig. 2.4. The dispersion relation described by Eq. (2.6) is shown in Fig. 2.5.

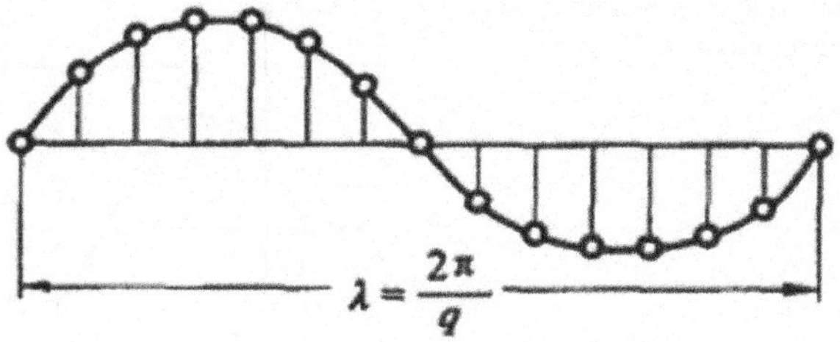

Fig. 2.4 Lattice wave in the crystal given by Eq. (2.5)

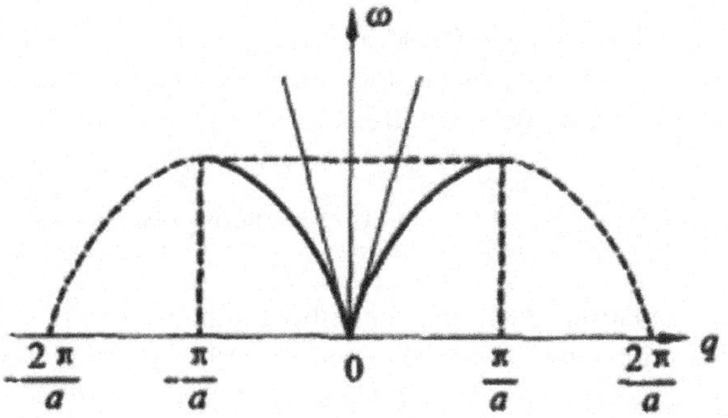

Fig. 2.5 Dispersion relation described by Eq. (2.6)

Clearly, the behavior of atomic vibration in one dimensional atomic chain is quite different from that of an atomic vibrator connected with a spring, the interaction between nearby atoms leads to the group exhibition occurring, which presents the rule of "1 + 1 > 2" in "systems theory", and a brand new exhibition that completely doesn't equal to any superposition of an atomic vibrator connected with a spring but reveals the wholeness of the physical system.

2.4 Travelling Waves and Standing Waves

In everyday life, we could see water wave and a beam of light, and hear acoustic waves and receive electromagnetic waves, which are all the transmission of vibration in spaces. The excitation source of the vibration is called wave origin (origin of vibration), the vibration of its surroundings is induced by the origin of vibration as a transmission of the vibration [1–6]. A wave at a certain instant in one dimension can be written as

$$\xi(x, t) = A\cos\left[\omega_0\left(t - \frac{x}{u}\right) + \varphi_0\right]. \tag{2.7}$$

In Eq. (2.7), A is the amplitude; φ_0 is the initial phase; ω_0 is its circular frequency; u is the transmission speed of wave in a medium (group speed); while $T_0 = 2\pi/\omega_0$ expresses the period of the vibration in time, and $\lambda = uT_0 = 2\pi u/\omega_0$ is wave length

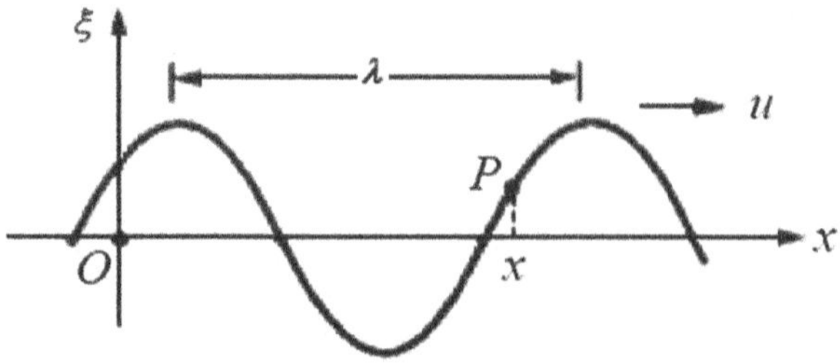

Fig. 2.6 Schematic of "right travelling wave"

that represents the period of wave spreading in space. Therefore, $\xi = \xi(x, t)$ reveals the time–space scene of the vibrations of wave origin and its surroundings globally.

An alternative form of Eq. (2.7) is,

$$\xi(x, t) = A\cos\left(\omega_0 t - \frac{2\pi x}{\lambda} + \varphi_0\right). \tag{2.8}$$

Equation (2.8) represents that the phase of the vibration along the transmission direction with a distance of wave length λ is falling—behind by 2π. The transmission direction of wave described by Eq. (2.8) is "right" one, so it is called "right travelling wave", as is shown in Fig. 2.6.

Analogically, the "left travelling wave" can be written as,

$$\xi(x, t) = A\cos\left[\omega_0\left(t + \frac{x}{u}\right) + \varphi_0\right] = A\cos\left(\omega_0 t + \frac{2\pi x}{\lambda} + \varphi_0\right). \tag{2.9}$$

Overall, despite of "right travelling wave" or "left travelling wave", all the points along the transmission routine in the transmission process of wave are connected, which move according to certain rule.

Furthermore, if the "right travelling wave" meets the "left travelling wave" in a space point, new phenomenon occurs. The space point will behave the superposition of both travelling waves.

Let $\xi_1(x, t) = A\cos(\omega_0 t - \frac{2\pi x}{\lambda} + \varphi_1)$ be the "right travelling wave", and $\xi_2(x, t) = A\cos(\omega_0 t + \frac{2\pi x}{\lambda} + \varphi_2)$ be the "left travelling wave", then the actual vibration of the space point is,

$$\xi(x, t) = \xi_1(x, t) + \xi_2(x, t) = A\cos\left(\omega_0 t - \frac{2\pi x}{\lambda} + \varphi_1\right) + A\cos\left(\omega_0 t + \frac{2\pi x}{\lambda} + \varphi_2\right). \tag{2.10}$$

Thus a new vibration manner appears in space,

$$\xi(x, t) = \xi_1(x, t) + \xi_2(x, t) = 2A\cos\left(\frac{2\pi x}{\lambda} - \frac{\varphi_1 - \varphi_2}{2}\right) \cdot \cos\left(\omega_0 t + \frac{\varphi_1 + \varphi_2}{2}\right). \tag{2.11}$$

Let's introduce an ordinate transformation,

$$\begin{cases} x' = x + x_0, \\ t' = t + t_0, \end{cases} \text{with} \begin{cases} \frac{2\pi x_0}{\lambda} = -\frac{\varphi_1 - \varphi_2}{2} \\ \omega t_0 = \frac{\varphi_1 + \varphi_2}{2} \end{cases}. \tag{2.12}$$

Then it derives

$$\xi(x', t') = 2A\cos\left(\frac{2\pi x'}{\lambda}\right) \cdot \cos(\omega_0 t'). \tag{2.13}$$

Equation (2.13) indicates that the vibration of point x' at time t' is $\cos(\omega_0 t')$ with a quasi–amplitude of $|2A\cos(2\pi x'/\lambda)|$. At the instant of $t' = 0$, the spatial configuration of wave is described by curve 1 in Fig. 2.7, till $t' = T/2$ the spatial configuration of wave is changed to be curve 5 in Fig. 2.7, others represents the intermediate states. In this case the vibration at each point keeps its vibration status instead of sole "right travelling" or "right travelling", which is called "standing wave".

The points with zero amplitude and maximum amplitude are named node and antinode (wave loop) of the standing wave, respectively. The distance between two nearby nodes or two nearby antinodes is λ/2. Again, this displays a group or collective phenomenon due to the interactions or synergistic effect among points in the system.

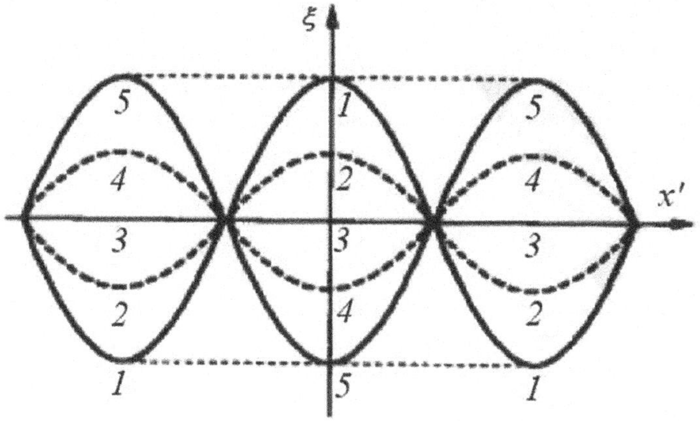

Fig. 2.7 Spatial configuration of wave of standing wave at different moment

2.5 Effect of Nano-size on Vibration of One Dimensional Atomic Chain

Suppose that the total number of atoms on a long chain is limited [11], says, N for example. Let l denote the order of atom on the chain, which takes values of 0, $\pm$ 1, $\pm$ 2, $\pm$ 3, ..., $(N-1)/2$; assume M express mass of the atom. The long chain consists of several segments of atomic chain, and the interfaces of between nearby segments are positioned at $l = 0$ and $l = 1$, $l = \pm n$ and $l = \pm n + 1$, $l = \pm 2n$ and $l = \pm 2n + 1$, ..., i.e., each segment contains n atoms. Let f express the force coefficient between two atoms within a segment, and f' present the force coefficient between two atoms on the boundaries of two neighbor segments, thus it satisfies $f' < f$, see Fig. 2.8.

Furthermore, introduce a parameter ε by $\varepsilon \equiv (f - f')/f$, then it leads to a result of $0 < \varepsilon \leq 1$ inevitably. Thus, the vibration frequency ω of atom on the one dimensional atomic chain becomes [11],

$$\omega^2 = \omega_k^2\left[1 - \frac{\varepsilon}{N}\left(\frac{N-1}{n} + 1\right)\right]. \tag{2.14}$$

Since $N \gg n \gg 1$, Eq. (2.14) further turns into,

$$\omega^2 \approx \omega_k^2\left[1 - \frac{\varepsilon}{n}\right] = \omega_k^2\left[1 - \frac{\alpha}{d}\right] < \omega_k^2. \tag{2.15}$$

In Eqs. (2.14) and (2.15), $d = na$ expresses the length (size) of atomic segment with a being separation of two neighbor atoms; $\alpha = \varepsilon a$ expresses the alternative coefficient of ε; ω_k is the vibration frequency of one-dimensional crystal with infinite length,

$$\omega_k^2 = \omega_M^2 \cdot \sin^2\left(\frac{ka}{2}\right), \omega_M^2 = \frac{4f}{M}. \tag{2.16}$$

Equation (2.15) indicates the decrease (red-shift) of atom vibration frequencies with respect of reduction of segment length in nano-size materials, of which the behaviors of the all atoms play their roles as a whole obviously. The red-shift effect

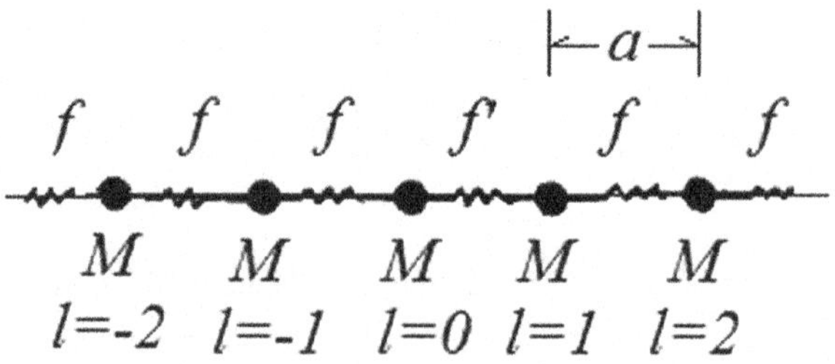

Fig. 2.8 Nano-size one dimensional atomic chain

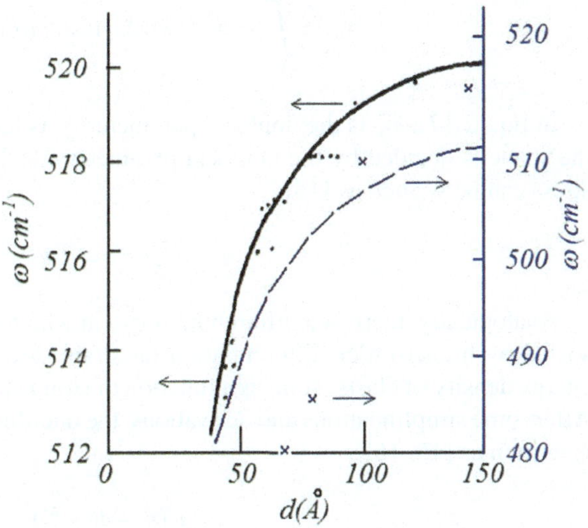

Fig. 2.9 Red-shift effect of atom vibration frequencies in nano-size materials. Symbols "·" and " × " indicate experimental data; solid and dotted lines indicate predictions from Eq. (2.15)

of atom vibration frequencies in nano-size materials is shown in Fig. 2.9 [11], which has been found in many experiments [11–13].

2.6 Nonlocal Continuum Mechanics

The generalized (or extended) continua include nonlocal theories as well as higher-order (or micro continuum) and higher-grade (or weak nonlocal) extensions to the classical continuum theories. Within nonlocal elasticity, nonlocal stress at a material point is expressed as a function of weighted values of the *entire stress field* [14]. Micro continuum is in fact *a continuous collection of deformable point particles*, for which the stress measurement includes *both force stress and couple stress*. The theory is in fact a two-level continuum model, in which macroscopic deformation field is enhanced with the microscopic deformation [14].

The nonlocal theory is dated back to 1960s, which aimed to *explain the experimental observations of the dispersion of elastic waves in the atomic lattices by using the hypothesis of the long-range forces* [15, 16]. *Classical local elasticity failed to depict this phenomenon.* While by pulling-in idea of the nonlocal parameter, the nonlocal theory is proposed to depict the phenomena including the dispersion of elastic waves in lattices, the homogenization of composite, the behaviors of nematic elastomers and carbon nanotubes [15].

In Eringen's nonlocal theory, *the stress at one point does not only depend on the strain at the appointed point but also on strains of all correlated points inside the body* [15], i.e.,

$$\sigma_{ij} = \int_{\Omega} \gamma(|x' - x|, \delta_{eg}) C_{ijkl} \varepsilon_{kl}(x') dv(x'). \tag{2.17}$$

In Eq. (2.17), δ_{eg} is the nonlocal parameter; γ is the nonlocal modulus, of which the shape is decided by the nonlocal parameter δ_{eg}. The one-dimensional nonlocal stress can be written as [15],

$$\sigma - \delta_{eg}^2 \sigma^{(2)} = E\varepsilon. \tag{2.18}$$

Analogically, there is a differential form, in which the nonlocal stress was represented by higher-order differentiation on strain. Mindlin formulated the potential energy density of elastic solid as a function of strain and its first and second gradient. After some simplifications and derivations, the one-dimensional stress strain relation is written as [15, 16],

$$\sigma = E(\varepsilon - \delta_{sg}^2 \varepsilon^{(2)}). \tag{2.19}$$

In Eq. (2.17), δ_{sg} is another nonlocal parameter.

In the damage mechanics based approach for fracture of metallic components, the stretched zone size (SZW) was proposed as the characteristic scale in the study [14]. Steel StE 690 was used to perform the tests at Aachen Technical University with specimens of size of 13 × 35 × 200 mm and the relative crack length a/W of 0.65, 0.75 and 0.85. The stretched zone is a typical area on the fracture surfaces, which can be determined to be about 54.6 μm by using fractographic analysis or scanning electron microscope after mechanical experiments [17].

In accordance with the viewpoint of physicist [17], *the formations of slip-bands and dislocation patterning are the consequences of highly correlated movements of dislocation group, the cooperative movement of dislocation groups is spreading over this characteristic scale.* Since the tested stretched zone size for StE steel is about 54.6 μm, about twice of the SZW of 125 μm was taken as the finite element mesh size around the crack tip to perform the damage evolution study. This choice for the element size means that one Gauss point describes in the representative manner the phenomenon of the highly correlated movement (slip-band) in the corresponding area at the crack tip within the framework of continuum plasticity. In some sense, it agrees with the viewpoint of physicist for presenting the behavior of slip-band at the crack tip.

It is *unnecessary to choose finer element size in the finite element mesh for classical continuum theories in a sense, so as to assure one Gauss point and its surrounding area big enough to cover the area of the stretched zone and to reveal the actual collective phenomena and interactions within it.* Or else, *if finer element size in the finite element mesh is selected*, the classical mechanics theory that is a macroscopic approach for continuum medium especially is used in a relatively microscopic scale in a sense, *it leads to the irrational results due to the overusing out of its valid range. For continuum mechanics*, it is valid for medium with homogeneity and even isotropicity, *which means that the valid scale should be beyond the inhomogeneity of the material studied*, i.e., *one should study the thing in a larger size which could cover the inhomogeneity within it and make the inhomogeneity indistinguishable.* The gross properties in the detected scale are homogeneous and even isotropic, so that an effective medium model (approximation) could be used reasonably.

2.7 Two-Body in Potential Field Problem

Two particles which are interacted each other could consist a closed system, the so called two-body system. The study of movement in two-body system is called two-body problem, which has analytical solution [18]. Especially in bound state, such as the moon running around the earth, the earth runs around the sun, the distance between the two particles keeps finite instead of infinite value.

Assume that the masses of the two particles (bodies) are m_1 and m_2 at the position coordinates $\overrightarrow{x_1}$ and $\overrightarrow{x_2}$, respectively, then the total kinetic energy of the two-body system is [18],

$$T = \frac{m_1}{2}\dot{\vec{x}}_1^2 + \frac{m_2}{2}\dot{\vec{x}}_2^2. \tag{2.20}$$

While the interaction potential energy is $V = V(\vec{x}_1, \vec{x}_2)$, which depends upon the relative position in space, and especially if the space is homogeneous and isotropic, then, it derives,

$$V = V(\vec{x}_1, \vec{x}_2) = V(|\vec{x}_1 - \vec{x}_2|). \tag{2.21}$$

Finally, the Lagrangian of the two-body system is,

$$L = \frac{m_1}{2}\dot{\vec{x}}_1^2 + \frac{m_2}{2}\dot{\vec{x}}_2^2 - V(|\vec{x}_1 - \vec{x}_2|). \tag{2.22}$$

In Eq. (2.22), the existence of interaction potential energy $V = V(|\vec{x}_1 - \vec{x}_2|)$ makes the movement of particle 1 connected with the information of particle 2, and vice versa. It is the so called "coupled".

There is approach to decouple the above "coupled" movements by introducing linear coordinate transformations [18], i.e.,

$$\begin{cases} \vec{x}_c = \dfrac{m_1\vec{x}_1 + m_2\vec{x}_2}{m_1 + m_2} \\ \vec{r} = \vec{x}_1 - \vec{x}_2 \end{cases}, \quad \begin{cases} m_r \equiv \dfrac{m_1 \cdot m_2}{m_1 + m_2} \\ m_c \equiv m_1 + m_2 \end{cases}. \tag{2.23}$$

Then it derives,

$$L_c = \frac{m_c}{2}\dot{x}_c^2, L_r = \frac{m_r}{2}\dot{r}^2 - V(r). \quad (2.24)$$

The Lagrangian L_c in Eq. (2.24) describes the movement of the *mass center of the two bodies with total mass of* m_c, which corresponds to the movement of the free particle with a mass m_c. While the Lagrangian L_r in Eq. (2.24) presents the *relative movement of the two bodies with the reduced mass of* m_r, which corresponds to the movement of a particle with the mass m_r in a central potential field.

As to the movement of a particle with the mass m in a central potential field, its Lagrangian in spherical coordinate system can be written as,

$$L = \frac{m}{2}(\dot{r}^2 + r^2\dot{\theta}^2 + r^2 \sin^2\theta\dot{\varphi}^2) - V(r). \quad (2.25)$$

The solution of Eq. (2.25) could describe the movement characteristics of celestial body.

As to "three-body problem", it is not too simple as those of two-body, complex phenomena might occur [19, 20].

2.8 Certainty and Chaos

1. **Deterministic principle**

The basic equations of Newtonian mechanics says that when the parametric conditions of the initial state are given, we can deduce the motion state of objects at all times after or before according to the basic equations, that is, these natural laws are deterministic and time symmetric [20]. There is no random component in certainty. For the deterministic model, as long as the input variables are set, the output is then deterministic reasonably. Another concept similar to certainty is determinism. Determinism is also called "Laplace creed". French mathematician Laplace once conceived a kind of "Laplace demon" on the basis of Newtonian mechanics. After knowing the exact position and momentum of each body in the universe, its present, past and future in the universe can be determined by using Newton's law. However, the next section will see the dramatic changes.

2. **Occurrence of chaos**

The opposite of "certainty" is "chaos" [20]. The modern connotation of chaos involves a wide range, it refers to a time-related process that is "random" in general, including fluctuation, bifurcation and uncertainty, etc. Turbulence is a typical chaotic phenomenon.

One source of chaos is the incompleteness of classical mechanical theory in describing multi-body or system. Newtonian mechanics and relativistic mechanics both focus on the description of the force between two objects (celestial bodies), but if the degree of freedom of the system increases, they will be stretched to record the state of the system accurately. The three-body problem (that is, the motion of a system containing three celestial bodies under the action of universal gravitation) has not been solved accurately until today. In addition, the basic equations in classical mechanics can not describe the irreversibility of time when describing the laws related to time, and the forward and reverse time will give the same law of motion. But when individuals are placed in the system, the interaction between them will bring entropy increase to the system, and its expression form is the second law of thermodynamics, that is, the irreversibility of time. In addition, the statistical methods in equilibrium and non-equilibrium state in thermodynamics also contradict the deterministic description of "trajectory" in classical mechanics. These insurmountable gaps between classical mechanics and thermodynamics inevitably lead to the end of certainty and the emergence of chaos.

3. **The Butterfly Effect**

One of the representatives of chaos theory is the so-called "Butterfly Effect", which was named by modern meteorologist Edward Lorenz. Its popular saying is that "a butterfly in Brazil can cause a tornado in Texas when it flaps its wings". Lorenz simplified the effective figures of three kinds of input data from 6 digits after the decimal point to 3 digits in numerical calculation of hydrodynamic closed system with finite mass, but the calculation results were completely different. He realized that the cause of the problem was that when the computer iterated, the operation error of each step was amplified, thus completely changing the state of the system. The butterfly effect shows that some factors seeming insignificant at first may play a decisive role in the later stage of the process. Till now, the weather forecast can only provide a probability. The reason why it failed weather forecast lies in the butterfly effect of every tiny variable in the whole weather system [20].

A better understanding of chaos and uncertainty depends on a deeper study of science and philosophy. But uncertainty may be one of the essential attributes of this world.

2.9 Summary

Above examples demonstrate that the interactions or synergistic effect among points in the system could result in group or collective phenomenon in the system obviously, which presents the rule of "$1 + 1 > 2$" in "systems theory" and completely doesn't equal to any superposition of behaviors of single particles but reveals the wholeness of the physical system. A deeper study of science and philosophy will provide a better understanding of chaos and uncertainty.

References

1. L.D. Landau, E.M. Lifshitz, *Mechanics, in Course of Theoretical Physics*, vol. 01, 3rd edn. (Butterworth—Heinenann, Linacre House, Jordan Hill, Oxford, UK, 2000)
2. B. Liu, *Mechanics* (China University of Science and Technology Press, Hefei, 2013)
3. Y.S. Shu, *Mechanics* (Peking University Press, Beijing, China, 2005)
4. X.H. Zhong, Y.M. Zhou, *Mechanics*, 2nd edn. (Peking University Press, Beijing, China, 2010)
5. M.J. Benacquista, J.D. Romano, *Classical Mechanics* (Springer, Cham, Switzerland, 2018)
6. J. Awrejcewicz, *Classical Mechanics* (Springer, New York, USA, 2012)
7. H. Vančik, *Philosophy of Chemistry* (Springer, Cham, Switzerland, 2021)
8. F. Capra, *The Tao of Physics, an Exploration of the Parallels Between Modern Physics and Eastern Mysticism*, 5th edn. (Shambhala Publications Inc., Boston, USA, 2013)
9. J.D. Patterson, B.C. Bailey, *Solid-State Physics, Introduction to the Theory*, 3rd edn. (Springer, Cham, Switzerland, 2018)
10. J.J. Quinn, K.S. Yi, *Solid State Physics, Principles and Modern Applications*, 2nd edn. (Springer, Cham, Switzerland, 2018)
11. M. Zheng, Z. Jin, Atom vibration of one-dimensional nano-phase materials. Chin. Sci. Bull. **39**(5), 382–386 (1994)
12. A.A. Nuhu, B. Safaei, State-of-the-art of vibration analysis of small-sized structures by using nonclassical continuum theories of elasticity. Arch. Computat. Methods Eng. **29**, 4959–5147 (2022)
13. H. Deng, Y. He, X. Wei, W. Chen, Y. Liu, X. Chen, H. Zhou, Crystal structure, vibration spectra and microwave dielectric properties for scheelite structure $NaSrLa(WO_4)_3$ ceramics and adjustment of its τ_f value. J. Mater. Sci. Mater. Electron. **34**, 1998 (2023)
14. S.M. Mousavi, Singularity-free defect mechanics for polar media. Continuum Mech. Thermodyn. **31**, 1883–1909 (2019)
15. J. Chen, *Nonlocal Euler–Bernoulli Beam Theories, A Comparative Study* (Springer, Cham, Switzerland, 2021)
16. Gérard A. Maugin, *Non-Classical Continuum Mechanics, A Dictionary* (Springer, Singapore, 2017)
17. M. Zheng, U. Lauschke, M. Kuna, A damage mechanics based approach for fracture of metallic components. Comput. Mater. Sci. **19**, 170–178 (2000)
18. X. Gao, *Classical Mechanics* (Science Press, Beijing, 2023)
19. P. Tan, X. Shen, X. Hou, X. Liao, A review on co-orbital motion in restricted and planetary three-body problem. Prog. Astron. **40**(1), 58–95 (2022)
20. W. Yang, P. Zhao, H. Wang, *Introduction to Mechanics* (Science Press, Beijing, 2020)

Chapter 3
Extreme Value Problems

Abstract The extreme value problems in natural world are quite common, such as distribution of electric charge on a spherical shell, distribution of electric current in an electric circuit, ground state energy of hydrogen atom, energy method in elasto-plastic mechanics, least action principle, etc. In this chapter, we will show the consequences of extreme value problems being the basic physical rules in nature in viewpoint of physics, and the essence of extreme value problems spontaneously exhibiting group behavior of the system and the synergetic actions among all constituent parts.

Keywords Extreme value · Electric charge · Electric current · Ground state energy · Hydrogen atom · Energy method · Least action principle · Group behavior · System

3.1 Introduction

In mathematics, extremum principle refers to get the extreme value of a function within its definition domain. It can be seen as an extension of variational method of analytical mechanics. The extreme value problems in natural are quite usual, such as distribution of electric charge on a spherical shell, distribution of electric current in an electric circuit, ground state energy of hydrogen atom, etc. The essence of extreme value problems behind is needed to be digging, in this chapter we will try to study this issue and to find the consequences of extreme value problems in temporal or spatial scopes, which shows the collaboration and synergetic action among parts of a system distinctly [1–4].

M. Zheng and J. Yu, *Systems Theory for Engineering Practice*,
https://doi.org/10.1007/978-981-97-9342-6_3

3.2 Extreme Value Problems in Electricity

Here we would like to show three extreme value problems in electricity [1]. One is the distribution density of electric charge on a spherical shell, the second is the current distribution of electricity in an electric circuit, and the other is the ground state energy of hydrogen atom.

Example 1. Distribution Density of Electric Charge on a Spherical Shell There are two concentric metal spheres with radii r and R, respectively, which are connected by thin electric wires. It is known that the whole system has a charge Q. Subsequently, the problem is to find the charge carried by the inner sphere q under condition of electrostatic balance.

Solution

If the self-energies of the inner sphere and the outer sphere are W_Q and W_q, and the interaction energy is W_{Qq}, respectively. Then according to the symmetry analysis, it is known that the charge must be symmetrically distributed on the spheres. Therefore, the total electrostatic energy of the system at electrostatic equilibrium is,

$$\begin{aligned} W &= W_q + W_Q + W_{Qq} = \frac{1}{2}k\frac{q^2}{r} + \frac{1}{2}k\frac{(Q-q)^2}{R} + k\frac{q(Q-q)}{R} \\ &= \frac{1}{2}kq^2\left(\frac{1}{r} - \frac{1}{R}\right) + \frac{1}{2}k\frac{Q^2}{R}. \end{aligned} \tag{3.1}$$

The condition for getting extreme value of energy is,

$$dW/dq = kq\left(\frac{1}{r} - \frac{1}{R}\right) = 0. \tag{3.2}$$

Since it has a condition of $r \neq R$, thus Eq. (3.2) leads to a inevitable result of $q = 0$, this consequence agrees with that of electrostatics. Simultaneously, it leads to a result of $d^2W/dq^2 = k\left(\frac{1}{r} - \frac{1}{R}\right) > 0$, which implies that the consequence of $q = 0$ corresponds to the *minimum value of electrostatic energy*.

Example 2. Current Distribution of Electricity in an Electric Circuit In a direct current (DC) circuit, the real process will also follow the extreme value principle.

1. **A parallel circuit**

Primarily, suppose that R_1 and R_2 are resistors in a parallel circuit. Given that the total current in the circuit is I, then the problem is to find the ratio of the currents passing through the two resistors.

Solution

We could use Ohm's law to solve this problem commonly. Alternatively, here energetic method is used. In fact, the current is distributed in such a way that *the total power of Joule heat generated in the circuit takes minimum.* If the total power of Joule heat of the circuit is P, and the currents passing through the resistors R_1 and R_2 are I_1 and I_2, respectively, then,

$$P = I_1^2 R_1 + I_2^2 R_2 = I_1^2 R_1 + (I - I_1)^2 R_2. \tag{3.3}$$

From the extremum condition, it reads,

$$dP/dI_1 = 0, \text{ and } I_1/I_2 = I_1/(I - I_1) = R_2/R_1. \tag{3.4}$$

This result is the same as that from Ohm's law.
Simultaneously,

$$d^2P/dI_1^2 = 2(R_1 + R_2) > 0. \tag{3.5}$$

Equation (3.5) implies that the consequence of $I_1/I_2 = R_2/R_1$ corresponds to the minimum total power of Joule heat generated in the circuit.

2. **A series circuit**

Next, consider a series circuit alternatively. It supposes that R_l and R_2 are resistors in a series circuit. If the total electric voltage in the circuit is given as U, then the problem is to find the ratio of the voltages getting in the two resistors.

Solution

In fact, the partial voltage is distributed in such a way that *the total power of Joule heat generated in the circuit takes minimum.* If the total power of Joule heat of the circuit is P and total voltage in the circuit is U, and the partial voltages distributed in the resistors R_1 and R_2 are U_1 and U_2, respectively, then,

$$P = U_1^2/R_1 + U_2^2/R_2 = U_1^2/R_1 + (U - U_1)^2/R_2. \tag{3.6}$$

Again from the extremum condition, it reads,

$$dP/dU_1 = 0,\ U_1 = UR_1/(R_1 + R_2) \text{ or } U_1/R_1 = U_2/R_2. \tag{3.7}$$

This result is the same as that from Ohm's law, too.
At the same time,

$$d^2P/dU_1^2 = 2(1/R_1 + 1/R_2) > 0. \tag{3.8}$$

Equation (3.8) indicates that the consequence of $U_1/R_1 = U_2/R_2$ is corresponding to the minimum total power of Joule heat generated in the circuit again.

Example 3. Ground State Energy of Hydrogen Atom Ground state energy of hydrogen atom is taken as the third example.

Let the radius of the electron "orbit" be a, which is moving around the nucleus (coordinate origin) and the momentum of electron be p, then it has a result of $p \sim \hbar/a$ due to uncertainty principle in quantum mechanics, $\hbar$ is Plank constant.

Solution

The kinetic energy of the electron of the hydrogen atom system is

$$E_k = \frac{1}{2}mv^2 = \frac{p^2}{2m} = \frac{\hbar^2}{2ma^2}. \tag{3.9}$$

While the electrostatic potential energy between the electron and the nuclear is

$$E_p = -k\frac{e^2}{a}. \tag{3.10}$$

Thus the total energy of the hydrogen atom system is,

$$E_T = E_k + E_p = \frac{\hbar^2}{2ma^2} - k\frac{e^2}{a}. \tag{3.11}$$

From the extremum condition of total energy [1], it reads,

$$dE_T/da = -\frac{\hbar^2}{ma^3} + k\frac{e^2}{a^2} = 0. \tag{3.12}$$

Equation (3.12) leads to the following consequences,

$$a = a_0 \equiv \frac{\hbar^2}{mke^2} = 0.53\,\text{A, and } E_T = -13.6\,\text{eV}. \tag{3.13}$$

This result is the same as that from exact quantum mechanics.
Simultaneously,

$$d^2E_T/da^2 = \frac{ke^2}{a^3} > 0. \tag{3.14}$$

It can be seen from Eqs. (3.12) and (3.14) that the action between electron and its nuclear in hydrogen atom is *to minimize the total energy of the atomic system* in ground state, which can be one of the important principles for natural arrangement.

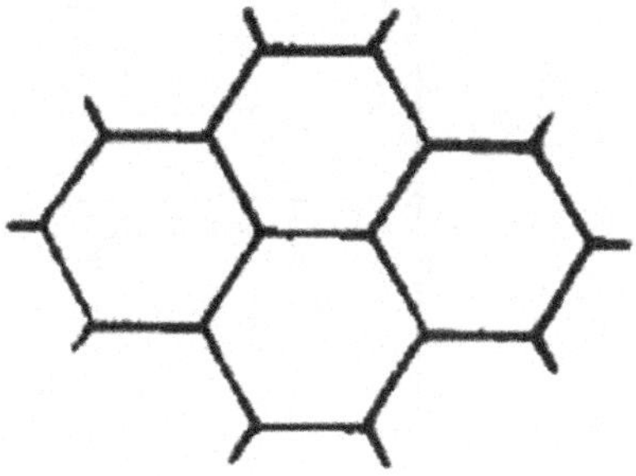

Fig. 3.1 Paving with regular hexagonal graphics

3.3 The Regular Hexagonal Structure of the Hive

As far as plane graphics are concerned, it is made up of many nearly regular hexagonal graphics, as is shown in Fig. 3.1. Why so?

This is because in the method of dividing a plane into parts with equal area, the regular hexagonal pattern frame grid has the smallest total grid length, that is to say, the occupation of materials is the least. In a plane figure with a certain perimeter, the area of a circle is the largest, but the circles can't fill the plane, and there will always be a gap between adjacent circles, so it is impossible to lay with circles [1]. Besides, there are only three kinds of regular polygons that can completely cover the plane: regular triangle, regular quadrangle and regular hexagon, and the regular hexagon has the smallest total length of the grid, i.e., the least material consumption of the system. Therefore, hive forms as that.

3.4 Energy Method

1. Application of principle of minimum total potential energy

Under conditions of certain actions of external forces and requirements of boundary condition, an elastic body keeps its equivalent state in a specific case that has a set of displacements corresponding to the minimum of total potential energy as a whole [2].

Figure 3.2 shows a cantilever beam, of which an external force P is applied on the free end. The length of this cantilever beam is l, its stiffness is EI. The problem is to find its maximum deflection of the beam.

Solution

Assume that the deflection function of this cantilever beam is

$$w = b_1 x^2 + b_2 x^3. \tag{3.15}$$

In Eq. (3.15), b_1 and b_2 are uncertain constants to be determined, which meet the following boundary requirements,

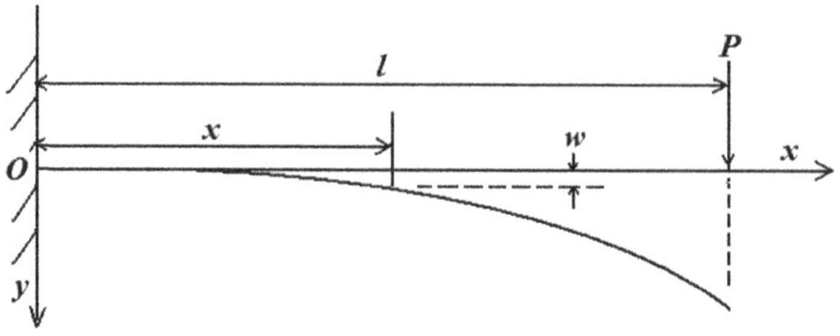

Fig. 3.2 A cantilever beam applied to an external force P at its free end

$$w|_{x=0} = 0,\ (\partial w/\partial x)|_{x=0} = 0. \tag{3.16}$$

The bending moment of the beam reads,

$$M(x) = -EI\frac{d^2 w(x)}{dx^2} = -EI(2b_1 + 6b_2 x). \tag{3.17}$$

Thus the strain energy of the bending beam is,

$$E_\varepsilon = \frac{1}{2EI}\int_0^l M^2(x)\cdot dx = \frac{EI}{2}\int_0^l (2b_1 + 6b_2 x)^2\cdot dx = 2EI(b_1^2 l + 3b_1 b_2 l^2 + 3b_2^2 l^3). \tag{3.18}$$

The total potential energy of the bending beam is,

$$E_p = E_\varepsilon - P\cdot w(x)|_{x=l} = 2EI(b_1^2 l + 3b_1 b_2 l^2 + 3b_2^2 l^3) - P(b_1 l^2 + b_2 l^3). \tag{3.19}$$

It derives following result according to principle of minimum total potential energy, i.e., $\delta E_p = 0$,

$$\partial E_p/\partial b_1 = 0, 2EI(2b_1 l + 3b_2 l^2) - Pl^2 = 0. \tag{3.20}$$

$$\partial E_p/\partial b_2 = 0, 2EI(3b_1 l^2 + 6b_2 l^3) - Pl^3 = 0. \tag{3.21}$$

Finally, the solutions are,

$$b_1 = \frac{Pl}{2EI}, b_2 = -\frac{P}{6EI}, w(x) = \frac{Plx^2}{6EI}\cdot\left(3 - \frac{x}{l}\right), w_{\max} = w(x)|_{x=l} = \frac{Plx^2}{3EI}. \tag{3.22}$$

The consequences are the same as those of the precise solutions [2].

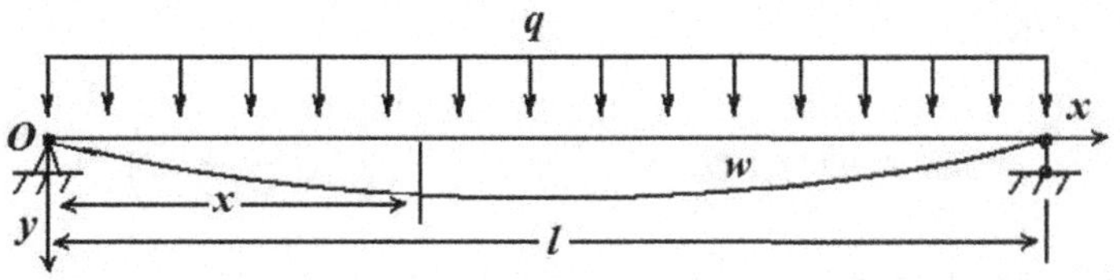

Fig. 3.3 A simply supported beam applied to a uniformly distributed load q

This example shows that *the displacement within the cantilever beam at every point behaves coordinately each other so as to get the total potential energy minimum of the beam system as a whole synergetically.*

2. **Approximation based on displacement variational method, Ritz-Rayleigh approach**

Under conditions of certain actions of external forces and boundary requirements, on basis of displacement variational method of virtual displacement principle, an approximate solution for elastic mechanics problem can be obtained [2].

Figure 3.3 shows a simply supported beam, which bears a uniformly distributed load q. The length of this simply supported beam is l, its stiffness is EI. Find its maximum deflection $w(x)$ by using Ritz-Rayleigh approach.

Solution

Assume that the deflection function of this simply supported beam is

$$w(x) = \sum_{n=1}^{\infty} C_n \sin \frac{n\pi x}{l}. \tag{3.23}$$

In Eq. (3.23), C_n is uncertain constant to be determined, which meets the following boundary requirements,

$$w|_{x=0} = 0,\ w|_{x=l} = 0. \tag{3.24}$$

Thus the total potential energy of the beam is,

$$E_p = \frac{EI}{2} \cdot \int_0^l \left(\frac{d^2w}{dx^2}\right)^2 \cdot dx - \int_0^l (q \cdot w) \cdot dx. \tag{3.25}$$

Finally, it derives,

$$C_n = \frac{4ql^4}{EI\pi^5} \cdot \frac{1}{n^5},\ (n = 1, 3, 5, \ldots);\ w(x) = \frac{4ql^4}{EI\pi^5} \cdot \sum_{n=1,3,\ldots}^{\infty} \frac{1}{n^5} \sin \frac{n\pi x}{l}. \tag{3.26}$$

At last, it gives,

$$w_{\max} = w(x)|_{x=l/2} = \frac{4ql^4}{EI\pi^5} \cdot \left(1 - \frac{1}{3^5} + \frac{1}{5^5} - \ldots\right). \tag{3.27}$$

Remaining the 1st term only, it gets

$$w_{\max} \approx \frac{4ql^4}{EI\pi^5} = \frac{4ql^4}{76.5EI}. \tag{3.28}$$

The result of Eq. (3.28) is close to that of the precise solution, i.e., $w_{\max 0} = \frac{4ql^4}{76.8EI}$ tightly [2].

3.5 Least Action Principle

Least action principle is a basic and universal principle in the natural system, which greatly promotes the scientific development [5]. The evolution of nature, and even the evolution of life, basically follows this law. It is reflected in various types of macro phenomena (such as in mechanics, heat, electromagnetism, optics and so on). In the process of all practical phenomena, it always meets the equation. The general expression is,

$$\delta S = \delta \int_{t_1}^{t_2} L \cdot dl(t) = 0. \tag{3.29}$$

In the general form of principle of least action (Hamilton principle of least action) given by Eq. (3.29), $S = \int_{t_1}^{t_2} L \cdot dl(t)$ is called action (Hamilton action), $l(t)$ is a function of time, and in the simplest case, $l(t) = t$, and L is an action function (Lagrangian function). As to different types of phenomena, the action functions L varies correspondingly. Equation (3.29) shows that all kinds of macro phenomena are carried out along the path of taking the extreme value of the action quantity in the entire itinerary. In mechanics, there exists principle of least action (Hamilton principle) for conservative fields,

$$\delta S = \delta \int_{t_1}^{t_2} L \cdot dt = \delta \int_{t_1}^{t_2} (T - V) \cdot dt. \tag{3.30}$$

According to Eq. (3.30), it derives Lagrange equation directly,

$$\frac{d}{dt}\frac{\partial L}{\partial \dot{q}} - \frac{\partial L}{\partial q} = 0. \tag{3.31}$$

In Eq. (3.31), the generalized momentum is defined by $p \equiv \partial L/\partial \dot{q}$. For the simple case of $L = \frac{1}{2}mv^2 - V(x)$, it derives Newton's second law directly,

$$F = ma, F = -\partial V(x)/\partial x \tag{3.32}$$

For special relativistic mechanics, its action function is

$$L = -mc^2\sqrt{(1 - v^2/c^2)}, p = \frac{\partial L}{\partial v} = mv/\sqrt{(1 - v^2/c^2)}. \tag{3.33}$$

While in general relativity [6, 7], it derives motion equation of a particle from the least action principle $\delta L = -mc\delta \int ds = 0$,

$$\frac{d^2x^i}{ds^2} + \Gamma^i_{kl}\frac{dx^k}{ds}\frac{dx^l}{ds} = 0. \tag{3.34}$$

According to this principle, a particle moves in such a way that its geodesic curve takes an extreme value between two given points. In other words, it moves along a shortest line in four-dimension space, x^0, x^1, x^2, x^3. Because of the existence of gravitational field, the space is not Euclid's straight space, so the shortest line is not a straight line. In Eq. (3.34), ds is the invariant, $ds^2 = -g_{ik}dx^i dx^k$, g_{ik} is metric tensor, Γ^i_{kl} is the Christoffel symbol which is a function of the metric tensor to reflect acceleration of gravity.

For heat, electromagnetism, optics, quantum mechanics, ..., they all follow principle of least action, and there exist exact action function correspondingly [5].

1. **The brachistochrone problem**

Figure 3.4 shows two points with the position relationship that point A is higher than point B. The problem is to find a curve from A to B such that a body could move from A to B with least time in the entire itinerary [3].

Solution

Set up a coordinate system as shown in Fig. 3.4. Then, for a point $P(x, y)$, according to law of energy conservation, the moving speed v of body is,

$$v = \sqrt{2gy}. \tag{3.35}$$

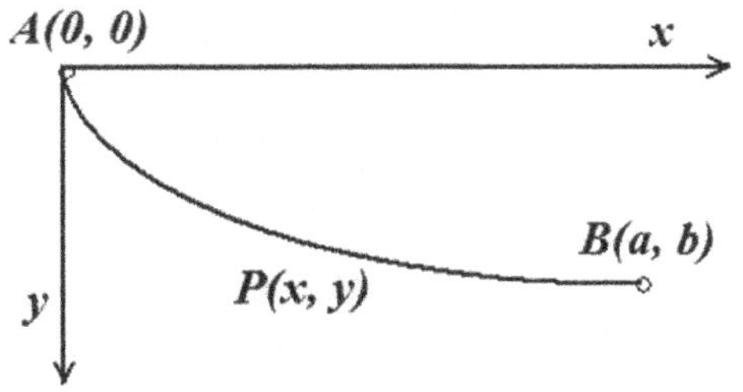

Fig. 3.4 Brachistochrone problem

In Eq. (3.35), g expresses gravitational acceleration.
Let dl be the arc length microelement, then it gets [3],

$$dl/dt \equiv v = \sqrt{2gy},\, dt = dl/v = dl/\sqrt{2gy} = \left[\sqrt{1+(dy/dx)^2}\cdot dx\right]/\sqrt{2gy}. \tag{3.36}$$

Thus the total time T for a body moving from point A to point B in the entire itinerary is,

$$T = \int_{x=0}^{a} \left[\sqrt{1+(dy/dx)^2}/\sqrt{2gy}\right]\cdot dx. \tag{3.37}$$

Furthermore, let $p = dy/dx$, then,

$$\begin{aligned} T &= \int_{x=0}^{a} \left[\sqrt{1+p^2}/\sqrt{2gy}\right]\cdot dx,\ \frac{\partial T}{\partial y} = -\frac{1}{\sqrt{2g}}\cdot\frac{\sqrt{1+p^2}}{2y^{3/2}},\ \frac{\partial T}{\partial p} \\ &= \frac{1}{\sqrt{2g}}\cdot\frac{p}{\sqrt{y}\cdot\sqrt{1+p^2}}. \end{aligned} \tag{3.38}$$

The corresponding Euler's equation reads,

$$\frac{\sqrt{1+p^2}}{2y^{3/2}} + \frac{d}{dx}\cdot\frac{p}{\sqrt{y}\cdot\sqrt{1+p^2}} = 0. \tag{3.39}$$

Equation (3.39) can be simplified as,

$$1 + \frac{2yp}{1+p^2}\cdot\frac{dp}{dy} = 0. \tag{3.40}$$

The integration of Eq. (3.40) leads to,

$$1 + p^2 = 1 + (dy/dx)^2 = 2C/y. \tag{3.41}$$

In Eq. (3.41), C is an integral constant. Further integration gives,

$$x = \int_{0}^{y} \left[(1/\sqrt{2Cy - y^2})\right]\cdot ydy = -\sqrt{2Cy - y^2} + C\cdot\cos^{-1}\left[(C-y)/C\right]. \tag{3.42}$$

Introduce a new parameter θ such that $\theta \equiv \cos^{-1}\left[(C-y)/C\right]$, then a parametric formula can be obtained [3],

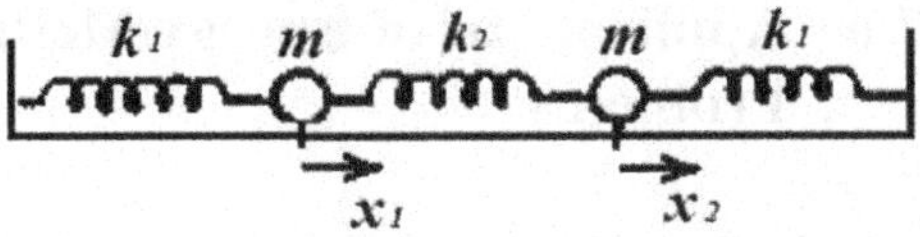

Fig. 3.5 A coupled-vibrator

$$x = C \cdot (\theta - \sin\theta),\ y = C \cdot (1 - \cos\theta). \tag{3.43}$$

Equation (3.43) is actually the parameter equation of cycloids, which gives the path curve from A to B with least time in the entire itinerary.

2. **Application of Lagrange equation in coupled-vibrator**

Shown in Fig. 3.5 is a coupled-vibrator, the masses of two bodies are the same, i.e., m, the stiffnesses of springs are k_1 and k_2, respectively. The problem is to find the vibration characteristics of the coupled-vibrator [4].

Solution
Select the equilibrium positions of two bodies as their coordinate origins, and let x_1 and x_2 be their instant displacements. Correspondingly, the total kinetic energy and total potential energy of the system can be written as,

$$E_k = \frac{1}{2}m\dot{x}_1^2 + \frac{1}{2}m\dot{x}_2^2,\ E_p = \frac{1}{2}k_1x_1^2 + \frac{1}{2}k_1x_2^2 + \frac{1}{2}k_2(x_2 - x_1)^2. \tag{3.44}$$

The action function (Lagrangian function) is $L = E_k - E_p$, then it derives following dynamic equations,

$$\begin{cases} m\ddot{x}_1 + k_1x_1 - k_2(x_2 - x_1) = 0, \\ m\ddot{x}_2 + k_1x_2 + k_2(x_2 - x_1) = 0. \end{cases} \tag{3.45}$$

Furthermore, let $q_1 = x_2 + x_1$, $q_2 = x_2 - x_1$, thus Eq. (3.45) results in,

$$\begin{cases} m\ddot{q}_1 + k_1q_1 = 0, \\ m\ddot{q}_2 + (k_1 + 2k_2)q_2 = 0. \end{cases} \tag{3.46}$$

The corresponding vibration circular frequencies are $\omega_1 = \sqrt{k_1/m}$ and $\omega_2 = \sqrt{(k_1 + 2k_2)/m}$, respectively. Obviously, the vibrations of body 1 and body 2 of the coupled-vibrator system are complex and coupled through circular frequencies ω_I and ω_2 and above coupling equations.

3.6 Applications of Energy Method in Solving Engineering Problems

1. Elastic limit of elliptical/circular pipes in condition of lateral force

Elastic limit analysis of an elliptical/circular pipe under lateral forces was developed by using energy method [8, 9], and analytical expression was given. Experimental results from literatures for compressing steel pipes and aluminum pipes were employed to check the proposed expression, satisfied results were obtained [8, 9].

Figure 3.6a shows a thin-walled elliptical pipe which is subjecting to a symmetric lateral tension.

Since the component and load are symmetric, a quarter of the component is analyzed, see Fig. 3.6b. The cross section *m–n* is shear-free and has a tensile force *P/2*, M_0 is the bending moment on the cross section *m–n* with the value to be undetermined.

Owing to the symmetry of the structure the cross section *m–n* is rotation—free, moreover the displacement corresponding to M_0 is zero [10], i.e.,

$$dU/dM_0 = 0. \tag{3.47}$$

In Eq. (3.47), U is the total strain energy of a quarter elliptical pipe in loading.

The bending moment of the cross section m_1–n_1 corresponding to the turning angle φ is,

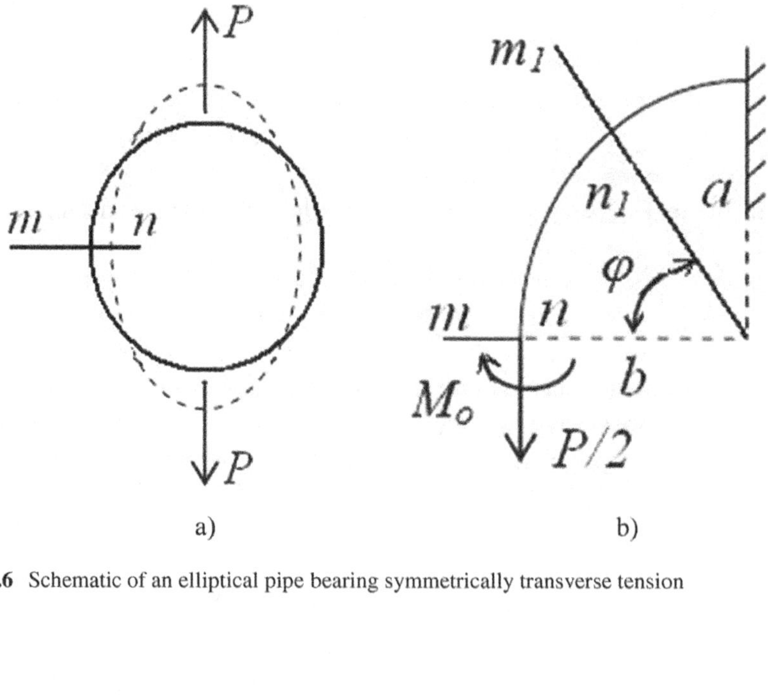

Fig. 3.6 Schematic of an elliptical pipe bearing symmetrically transverse tension

$$M = M_0 - \frac{P}{2}\mathrm{b}(1 - \cos\varphi). \tag{3.48}$$

In Eq. (3.48), b expresses the half length of the short axis of the ellipse, and,

$$dM/dM_0 = 1. \tag{3.49}$$

Meanwhile, the total potential energy of deformation is [10],

$$U = \int_0^S \frac{M^2}{2EI_Z}\mathrm{ds}. \tag{3.50}$$

In Eq. (3.50), E is the elastic modulus of the pipe material, I_z is the moment of inertia, and ds is the increment of the elliptical arc.

It further yields [8, 9],

$$\begin{aligned} 0 &= \int_0^{S_{\pi/2}} \left[M_0 - \frac{P}{2}\mathrm{b}(1 - \cos\varphi)\right] ds \\ &= \int_0^{\pi/2} \left[M_0 - \frac{P}{2}\mathrm{b}(1 - \cos\varphi)\right] \cdot \sqrt{a^2\cos^2\phi + b^2\sin^2\varphi} \cdot d\varphi. \end{aligned} \tag{3.51}$$

In Eq. (3.51), a expresses the half length of the long axis of the ellipse.

Moreover, two auxiliary parameters r and κ are introduced to characterize the shape of the ellipse,

$$a = r(1+\kappa), b = r(1-\kappa). \tag{3.52}$$

In Eq. (3.49), r is in fact the radius of the corresponding perfect circular pipe, and κ indicates the deviation degree of the instant elliptical pipe from the complete circular one, which reflects the variation of the deforming pipe and the ellipticity of the elliptical pipe. Parameter $\kappa = 0$ corresponds to the perfect circular pipe with radius r.

Finally, from Eq. (3.51), it derives,

$$M_0 = \frac{P \cdot r \cdot (1-\kappa)}{2} \cdot \left\{1 - \frac{1}{\pi\left(1+\kappa^2/4\right)} \cdot \left[1 - \varsigma + \frac{(1+\kappa)^2}{\sqrt{4\kappa}} \cdot \arcsin\left(\frac{\sqrt{4\kappa}}{1+\kappa}\right)\right]\right\}. \tag{3.53}$$

Moreover, it yields [6, 7],

$$
\begin{aligned}
M &= M_0 - \frac{P \cdot b}{2}(1 - \cos\varphi) \\
&= \frac{P \cdot r \cdot (1-\kappa)}{2} \cdot \left\{ \cos\varphi - \frac{1}{\pi\left(1+\kappa^2/4\right)} \cdot \left[1 - \kappa + \frac{(1+\kappa)^2}{\sqrt{4\kappa}} \cdot \arcsin\left(\frac{\sqrt{4\kappa}}{1+\kappa}\right) \right] \right\}.
\end{aligned} \tag{3.54}
$$

The maximum bending moment appears at the loading points $\varphi = \pi/2$ for a given loading P, i.e.,

$$
M_{\pi/2} = -\frac{P \cdot r \cdot (1-\kappa)}{2} \cdot \frac{1}{\pi\left(1+\kappa^2/4\right)} \cdot \left[1 - \kappa + \frac{(1+\kappa)^2}{\sqrt{4\kappa}} \cdot \arcsin\left(\frac{\sqrt{4\kappa}}{1+\kappa}\right) \right]. \tag{3.55}
$$

Furthermore, when the load P increases to its critical value P_c, plastic yielding occurs first at the loading point $\varphi = \pi/2$,

$$
P_c = \frac{2M_e}{r} \cdot \frac{\pi(1+\kappa^2/4)}{\left[1 - \kappa + \frac{(1+k)^2}{\sqrt{4\kappa}} \cdot \arcsin\left(\frac{\sqrt{4\kappa}}{1+\kappa}\right)\right] \cdot (1-\kappa)}. \tag{3.56}
$$

In Eq. (3.56), M_e presents the plastic bending moment of a sheet with the thickness t and width w [8, 9],

$$
M_{\mathrm{e}} = \frac{\sigma_s t^2 w}{4}. \tag{3.57}
$$

where σ_s is the plastic yielding strength of the pipe material. Here, elastic–perfect plastic material model is employed.

By substituting Eq. (3.51) into Eq. (3.56), it yields,

$$
P_c = \frac{\sigma_s t^2 w}{2r} \cdot \frac{\pi(1+\kappa^2/4)}{\left[1 - \kappa + \frac{(1+k)^2}{\sqrt{4\kappa}} \cdot \arcsin\left(\frac{\sqrt{4\kappa}}{1+\kappa}\right)\right] \cdot (1-\kappa)}. \tag{3.58}
$$

Equation (3.58) gives the estimation for critical load of an elliptical pipe under condition of lateral tension, which reveals that the value of P_c decreases with the reduction of κ.

From Eq. (3.58), for a perfect circular pipe, i.e., $\kappa = 0$, it yields the critical load for a circular pipe under condition of lateral tension,

$$
P_{Ce} = \frac{\pi \sigma_s t^2 w}{4r} \tag{3.59}
$$

It can be seen from Eq. (3.59) that the critical load value of the perfect circular pipe is not greater than the corresponding elliptical pipe under lateral tension. An

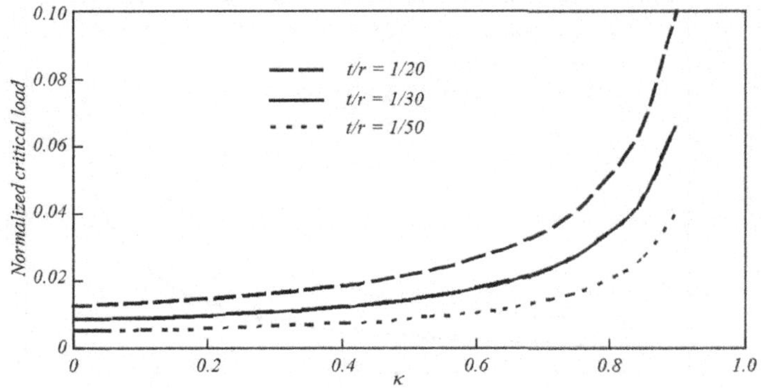

Fig. 3.7 Variation of the normalized critical load $P_c/[\sigma_s(t \cdot w)$ with respect to t/r and κ

alternative form of Eq. (3.58) is,

$$P_c/[\sigma_s(t \cdot w)] = \frac{\pi t}{2r} \frac{(1 + \kappa^2/4)}{\left[1 - \kappa + \frac{(1+\kappa)^2}{\sqrt{4\kappa}} \cdot \arcsin\left(\frac{\sqrt{4\kappa}}{1+\kappa}\right)\right] \cdot (1 - \kappa)}. \tag{3.60}$$

The term $P_c/[\sigma_s(t \cdot w)$ in Eq. (3.60) gives the normalized critical load of the elliptical pipe under transverse tension, which increases with the increase of the parameters t/r and κ tremendously, as is shown in Fig. 3.7.

Furthermore, by ignoring all minor effects, such as pipe length, friction and heating during the experiment, the actual critical load might be the same for pipes with the same instantaneous shape despite of loading type whether pulling or compressing.

Moreover, Gupta's experimental data can be used to check the estimation of the critical load of the elliptical pipe under lateral compression [8, 9, 11].

Table 3.1 cites the pipe parameters used in Gupta experiments and calculations [11].

Substituting the material parameters from Table 3.1 into Eq. (3.58), the estimations of critical loads for the A503 and S504 pipes under lateral compression can be obtained, Fig. 3.8 shows the comparison.

Similarly, Lipa's experimental data can be employed to check the estimation here once more. Table 3.2 cites the pipe parameters used in the Lipa's experiments [12].

Table 3.1 Material parameters of laterally compressive pipes A503 and S504 [11]

Specimen no.	A503	S504
Average radius (r) (mm)	23.83	22.4
Thickness(t) (mm)	3.44	3.34
r/t	6.925	7.120
Width (w) (mm)	100	100
Material properties σ_s (MPa)	153.3	288.2

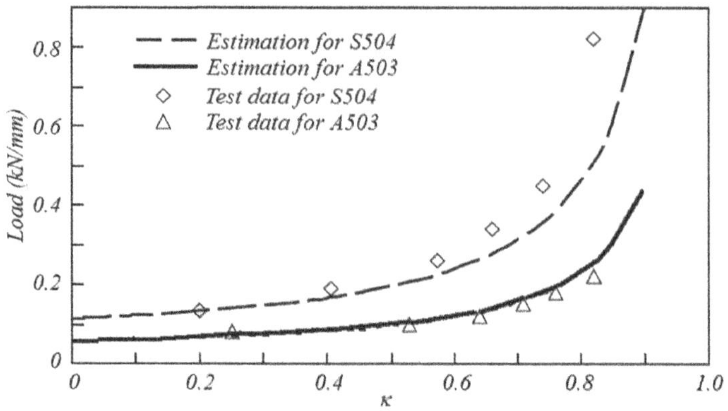

Fig. 3.8 Comparison of the estimation and laterally compressive tests for circular pipes A503 and S504

Figure 3.9 shows a comparison of the estimations of the Eq. (3.58) with the experimental results of Lipa and Kotełko, meanwhile the theoretical results of de Runtz and Hodge are presented for comparison as well. Distinctly, the elastic limit problem of elliptical/circular pipes in condition of lateral force, which involves complex deformation process, is solved by using energy method with principle of total energy extreme of the system properly.

Table 3.2 Tubular parameters of Lipa and Kotełko's experiments [12]

Thickness t (mm)	Material properties σ_s (MPa)	Tube diameter $2R$ (mm)	Width l (mm)	R/t
3.2	378	88.9	100	13.891

Fig. 3.9 Comparison of predictions of de Runtz and Hodge and Eq. (3.58) with Lipa and Kotełko's experimental results

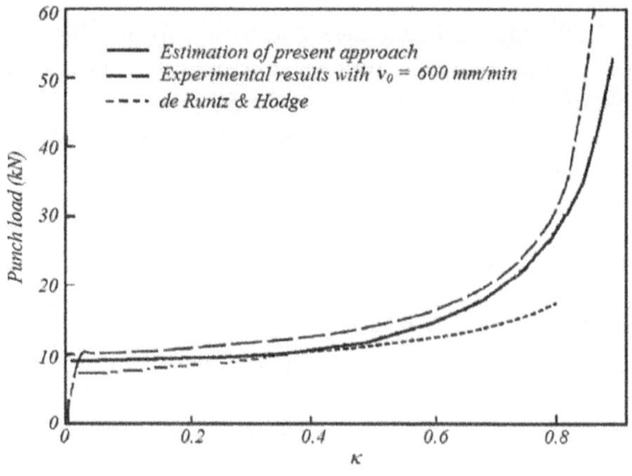

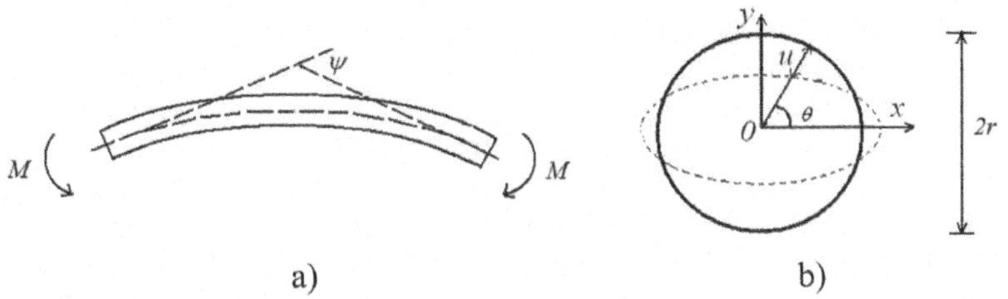

Fig. 3.10 Elliptical cross section model for pipe bending [8, 13, 14]

Figures 3.8 and 3.9 show that the estimations from Eq. (3.58) agree with the experimental data essentially.

2. **Plastic buckling during pipe bending**

2.1 **Elliptical cross section model for plastic bending pipe**

Considering that the static bending of a thin-wall circular pipe in its longitudinal direction by an instantaneous bending moment M, see Fig. 3.10a, the shape of cross section of the pipe changes correspondingly. Thus the longitudinal compressive and tensile stresses make the cross section ovalization [8, 13, 14], as is shown in Fig. 3.10b.

According to Brazier, the cross section changes to a typical elliptical shape, which can be expressed as [8, 13, 14],

$$u = r\gamma\cos(2\theta) \tag{3.61}$$

In Eq. (3.61), r indicates the average radius of the original tube; u is the radial displacement of pipe with respect to original position of a studied point; θ is the angle of a point in polar coordinate system; γ is a dimensionless coefficient which characterizes the radial displacement; see Fig. 3.10. An elliptical cross section model was proposed to analyze the change of cross section shape of the bending pipe [8, 13, 14], see Fig. 3.10.

2.2 **Energetic approach**

Furthermore, by using energy method, it obtained the total bending strain energy rate containing cross section ovalization of a pipe with length l is [8, 13, 14],

$$\dot{W} = \dot{W}_b + \dot{W}_{ovalization}, \tag{3.62}$$

i.e.,

$$\begin{aligned}\dot{W} = \frac{M_0 \cdot l\dot{C}}{\sqrt{2}} \cdot \Bigg\{ &\left(1 - \frac{2\gamma}{3} - \frac{\gamma^2}{3}\right) \\ &+\alpha \cdot \left[\frac{(3+\gamma)}{(1-\gamma)^3} \cdot (1-0.95\gamma)^{0.5} - \frac{(3-\gamma)}{(1+\gamma)^3} \cdot [2-(1-0.95\gamma)^{0.5}]\right]\Bigg\},\end{aligned} \tag{3.63}$$

in which

$$\alpha = \frac{\pi t}{32r^2}\frac{\dot{\gamma}}{\dot{C}} \tag{3.64}$$

In Eq. (3.64), α is the adjusted coefficient, which reflects the energetic relation between the pure bending and cross section shape changing of the bending tube, it is the ratio of the shape changing rate $\dot{\gamma}$ with respect to the longitudinal bending curvature rate $\dot{C}$.

Equation (3.63) indicates the dependence of energy change rate for a pipe bending on the parameter $\dot{C}$ and the adjusted coefficient α. In practice, α will self-adjust to make the total energy required for the pipe bending minimum of the system naturally, i.e.,

$$\frac{\partial \dot{W}}{\partial \gamma} = 0. \tag{3.65}$$

Moreover, it yielded the critically plastic buckling strain of the outer- fiber-line for the pipe bending through complicated mathematical derivations [8, 13],

$$\varepsilon_c = 0.19\frac{t}{r}\cdot\left(1+\frac{t}{1.78r}\right) \approx 0.19\frac{t}{r}. \tag{3.66}$$

The analytical factor 0.19 in Eq. (3.66) is close to the most experimental results [6, 11], it indicates the rationality of the energetic approach for this issue.

The comparison of the estimations of Eq. (3.66) with the experimental data is shown in Fig. 3.11. It can be seen from Fig. 3.11 that the present estimations agree with the available test data as a whole [8, 13].

Furthermore, in considering strain-hardening effect for plastic bending pipe at buckling, it derives [15],

$$\varepsilon_c = 0.19\frac{t}{r}\cdot\left(1+\frac{t}{1.78r}\right)\cdot[0.096 + 0.904/(1-n)^2]. \tag{3.67}$$

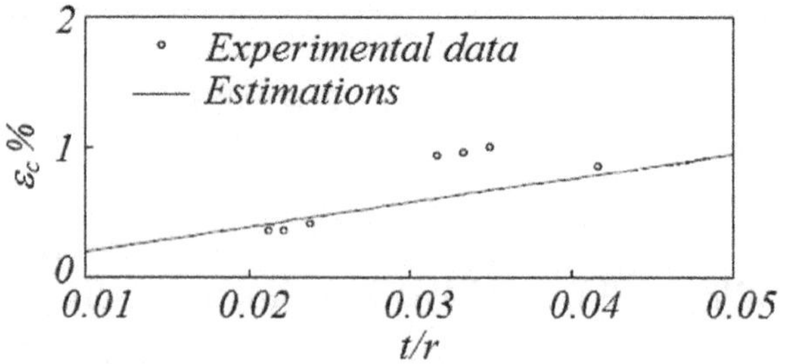

Fig. 3.11 Comparison of estimations with available experimental data

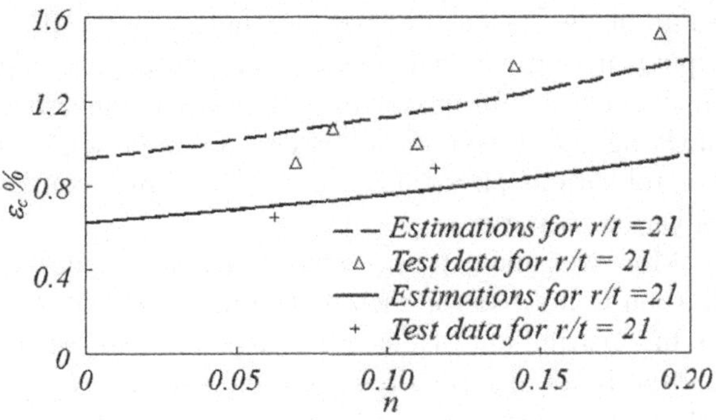

Fig. 3.12 ε_c versus n for Hollomon type of pipes with varying r/t

In Eq. (3.67), n indicates the exponent of strain hardening of the pipe material in Hollomon formula, $\sigma = k\varepsilon^n$ [8, 15].

Figure 3.12 shows the comparison of the estimations of Eq. (3.67) with the experimental data [8, 15]. Figure 3.12 clearly reveals the validity of the expression Eq. (3.67) for ε_c *vs* n for varying r/t value. Obviously, the synergistic action or cooperativity in a complex deforming system is displayed by using energy method, the mutual cooperativity of each material element performs under the constraint of principle of total energy extreme of the system voluntarily.

3.7 Discussion on Overall Approach and Its Applications

Extremum principle is a broad concept, which covers many disciplines, such as mathematics and physics. In mathematics, the extreme value principle usually refers to the maximum or minimum value of a function at a specific point, which reflects the maximum or minimum value of a function in its neighborhood. In the fields of physics and engineering, the extremum principle is also more widely used, such as in incremental analysis of elastoplastic mechanics [16], slope stability analysis in civil engineering [17], energy consumption of water flowing in hydraulics [18], and even analysis of animal behavior modeling [19], etc., in additional to the typical examples of this chapter. Other example, such as in the heat conduction equation, the extremum principle may involve the maximum or minimum value of temperature distribution, which is very important for understanding and predicting the behavior of physical phenomena. Besides, when solving practical problems, knowing the maximum or minimum temperature of an object can help optimize the heat dissipation, and improve efficiency or avoid overheating in design. In addition, the extremum principle also plays an important role in the optimal control theory, especially when dealing with the system optimization problem with constraints, the extremum principle provides an optimal control method to determine the maximum or minimum value of the system performance index. Generally speaking, the extremum principle is not only a mathematical concept, but also a powerful tool for solving optimization problems in systems of physics, engineering and other scientific fields. By studying and applying the extreme value principle, we can better understand and control various physical processes, thus improving the efficiency and performance of various systems.

Moreover, the extreme value principle in nature exhibits some balance or equilibrium in a sense for each part in the aspect of its contribution or function, such as the distribution of electric charge on a spherical shell, distribution of electric current in an electric circuit, ground state energy of hydrogen atom, etc., it actually induces extreme value for energy in total for a system, so it could be a quite usual principle in the world. Energy method, least action principle, etc., might be the representative of extremum principle for energy of a system as a whole. In many disciplines extremum principle has been widely used, fantastic achievements are obtained, which evidences the universality and validity of extremum principle in a system. The examples presented in this chapter are only short parts of the practices, the collaboration and synergetic action among constituent parts of the systems are quite common, the balance or equilibrium for each part in the aspect of its contribution or function constitute the extremum principle of a system for energy entirely. This might be the natural rule of harmonious coexistence of constituting parts in a system, i.e., balance or equilibrium for each part in the aspect of its contribution or function, extreme value for a system as an ensemble. This is the real scene of "competition and harmony coexistence" in a system.

3.8 Summary

The movements of many bodies are connected and synergetic in a system either in temporal or spatial scopes; Extreme value problems become inevitable routine to control the processes of the system as a whole; Collaboration and synergetic action among parts of the system are beyond any superposition of single parts.

References

1. L. Xu, *Principle of Least Action and the Development of Physics* (Sichuan Educational Press, Chengdu, 2001)
2. D.F. Sheng, M.B. Li, D.B. Zhu, *Elastoplastic Mechanics* (Science Press, Beijing, 2022)
3. H.C. Hu, *Variational Principle and its Application of Elastoplastic Mechanics* (Science Press, Beijing, 1981)
4. X.H. Zhong, Y.M. Zhou, *Mechanics*, 2nd edn. (Peking University Press, Beijing, China, 2010)
5. D.W. Liu, On the principle of least action. Sci. Technol. Eng. **7**(10), 2319–2320 (2007)
6. Ø. Grøn, *Introduction to Einstein's Theory of Relativity, From Newton's Attractive Gravity to the Repulsive Gravity of Vacuum Energy*, 2nd edn. (Springer, Cham Switzerland, 2020)

7. P. Fleury, *Gravitation, From Newton to Einstein* (Springer, Cham, Switzerland, 2019)
8. M. Zheng, Z. Yin, H. Teng, J. Liu, Y. Wang, *Elastoplastic Behavior of Highly Ductile Materials* (Springer, Singapore, 2019)
9. M. Zheng, Y. Zhao, H. Teng, J. Hu, L. Yu, Elastic limit analysis for elliptical and circular tubes under lateral tension. Arab. J. Sci. Eng. **40**, 1727–1732 (2015)
10. S. Timoshenko, *Strength of Materials, Part 1, Elementary Theory and Problems*, 3rd edn. (Stanford University, Stanford, USA, 1955)
11. N.K. Gupta, G.S. Sekhon, P.K. Gupta, Study of lateral compression of round metallic tubes. Thin Walled Struct. **43**, 895–922 (2005)
12. S. Lipa, M. Kotełko, Lateral impact of tubular structure—theoretical and experimental analysis. Part 1: investigation of single tube. J. Theor. Appl. Mech. **51**(4), 873–882 (2013)
13. L.K. Ji, M. Zheng, H.Y. Chen, Y. Zhao, L.J. Yu, J. Hu, H.P. Teng, Apparent strain of a pipe at plastic bending buckling state. J. Braz. Soc. Mech. Sci. Eng. **37**, 1811–1818 (2015)
14. T. Wierzbicki, M.V. Sinmao, A simplified model of Brazier effect in plastic bending of cylindrical tubes. Int. J. Pres. Ves. Pip. **71**, 19–28 (1997)
15. M. Zheng, Z.M. Li, Strain–hardening effect on critical strain assessment of pipe plastic bending at buckling. Mater. Phys. Mech. **44**, 48–60 (2020)
16. S. Du, Incremental analysis of elasto-plastic material and extremum theorems. Acta Armamentarrii **8**, 56–59 (1987)
17. W. Zhang, J. Wen, C. Ming, Imbalance trust force method for slope satibility analysis based on extemum principle. Soil Eng. Found. **23**(4), 81–84 (2009)
18. C. Huang, B. Xi, Study on extreme theories of energy consumption and its hydraulics examples. J. Yangtze River Sci. Res. Inst. **19**(5), 7–9 (2002)
19. X. Liang, Application of extreme principle of function in analysis of animal behavior modeling. J. Math. Technol. **16**(3), 17–21 (2000)

Chapter 4
Characteristics of Electron in Atom, Molecule and Metal

Abstract The characteristics of electrons in atom, molecule and metals are comparatively outlined, which include the characteristics of electrons in hydrogen atom, helium atom and hydrogen molecule, status of electrons in atoms and metals, first ionization energy of atoms and work function of metals, and electrical resistivity of electrons in metals. The differences of characteristics of electrons in above cases reveal the group behavior of the electronic systems and their synergetic actions among all electrons; the electrical resistivity of electrons in metals reflects the interaction between electrons and crystal lattices and defects.

Keywords Electrons · Atom · Metals · First ionization energy · Work function · Electrical resistivity · Group behavior

4.1 Introduction

In nature, materials exist in different forms, and the group effects exist almost in their statuses everywhere. Here in this chapter, the energies to drag an electron from hydrogen atom, helium atom, hydrogen molecule, and bulk metals are shown to reveal the group effects and interactions of electrons in the corresponding systems.

4.2 Electrons in Hydrogen and Helium Atoms

1. **Characteristic of electron in hydrogen atom**

Hydrogen atom is the simplest atom, which has only one electron moving in the electrostatic field (Coulomb field) of the nucleus of one proton, $V(r) = -\frac{1}{4\pi\varepsilon_0} \cdot \frac{Ze^2}{r}$, $Z = 1$ for hydrogen. Take the position of nucleus as the coordinate origin of a spherical coordinate system, the static wave function $u(r, \theta, \varphi)$ can be assumed, which meets following Schrodinger's equation in stationary state [1–6],

M. Zheng and J. Yu, *Systems Theory for Engineering Practice*,
https://doi.org/10.1007/978-981-97-9342-6_4

$$\nabla^2 u + \frac{2m_e}{\hbar^2}\left(E + \frac{Ze^2}{4\pi\varepsilon_0 r}\right)u = 0, \tag{4.1}$$

$$\frac{1}{r^2}\frac{\partial}{\partial r}\left(r^2\frac{\partial u}{\partial r}\right) + \frac{1}{r^2\sin\theta}\frac{\partial}{\partial r}\left(\sin\theta\frac{\partial u}{\partial\theta}\right) + \frac{1}{r^2\sin\theta}\frac{\partial^2 u}{\partial\phi^2} + \frac{2m}{\hbar^2}\left(E + \frac{Ze^2}{4\pi\varepsilon_0 r}\right)u = 0, \tag{4.2}$$

The exactly analytical solutions of Eq. (4.2) read,

$$E_n = -\frac{m_e e^4 Z^2}{2(4\pi\varepsilon_0)^2 n^2\hbar^2}, \quad n = 1, 2, 3 \ldots, \tag{4.3}$$

$$u(r, \theta, \varphi) = R(r)\Theta(\theta)\Phi(\varphi). \tag{4.4}$$

In Eq. (4.3), E_n is called energy level, which is quantified; $\Delta E_n = E_{n+1} - E_n$ is called energy gap between levels. While in Eq. (4.4), $R(r)$, $\Theta(\theta)$, and $\Phi(\varphi)$ are functions of only r, θ, and φ individually, which have specific expressions [1–6]. In Eq. (4.3), Z is the number of nucleus charge.

The remarkable features of above solution are that the ground energy of hydrogen atom and ground wave functions are,

$$E_1 = -\frac{m_e e^4 Z^2}{2(4\pi\varepsilon_0)^2\hbar^2} = -\frac{m_e e^4}{2(4\pi\varepsilon_0)^2\hbar^2} = -13.60 \text{ eV}, \tag{4.5}$$

$$R_{1,0}(r) = \left(\frac{Z}{a_0}\right)^{3/2} \cdot 2\exp\left(-\frac{Zr}{a_0}\right) = \left(\frac{1}{a_0}\right)^{3/2} \cdot 2\exp\left(-\frac{r}{a_0}\right). \tag{4.6}$$

The most important consequences include that the ionization energy of hydrogen atom is 13.60 eV, Rydberg constant $\text{Re} \equiv \frac{m_e e^4}{4\pi(4\pi\varepsilon_0)^2\hbar^3 c} = 10973731.1 \text{ m}^{-1}$ and the 1st Bohr radius of hydrogen atom $a_0 \equiv \frac{4\pi\varepsilon_0\hbar^2}{m_e e^2} = 5.2917721067 \times 10^{-11}$ m, which agree with experimental data. At the same time, $E_n = -\frac{m_e e^4 Z^2}{2(4\pi\varepsilon_0)^2 n^2\hbar^2} = \frac{e^2 Z^2}{8\pi\varepsilon_0 n^2 a_0}$.

2. **Characteristic of electrons in helium atom**

Helium atom has two electrons moving in the electric field (Coulomb field) of the nucleus of two protons, $V(r) = -\frac{1}{4\pi\varepsilon_0}\cdot\frac{Ze^2}{r}$, $Z = 2$ here.

Since the electronic number is 2, so helium is comparatively complex. Assume that the Hamiltonian of helium atom is,

$$\hat{H} = \hat{H}^{(0)} + \Delta\hat{H}, \tag{4.7}$$

$$\hat{H}^{(0)} = -\frac{\hbar^2}{2m_e}\nabla_1^2 - \frac{2e^2}{4\pi\varepsilon_0 r_1} - \frac{\hbar^2}{2m_e}\nabla_2^2 - \frac{2e^2}{4\pi\varepsilon_0 r_2}, \quad \Delta\hat{H} = \frac{e^2}{4\pi\varepsilon_0 r_{12}}. \tag{4.8}$$

In Eq. (4.8), r_{12} is the distance between two electrons, ∇_1^2 and ∇_2^2 express nabla symbols of the two electrons' positions of r_1 and r_2. The precise analytical solution of helium atom is impossible, so it could get an approximate solution only.

The $\Delta\widehat{H}$ expresses the direct electrostatic interaction; the Hamiltonian $\widehat{H}^{(0)}$ is a summation of two hydrogen-like atoms' Hamiltonians with $Z = 2$, so its eigen value $E_0^{(0)}$ is the summation of two hydrogen-like atoms' energies, eigen wave function is the product of two hydrogen-like atoms' wave functions. As to the ground state at zero-th order approximation [1–6],

$$u^{(0)}(r_1, r_2) = u_{100}(r_1) \cdot u_{100}(r_2) = \frac{8}{\pi a_0^3} e^{-\frac{2(r_1+r_2)}{a_0}}. \tag{4.9}$$

Thus the first-order correction to energy is,

$$\begin{aligned} E_0^{(1)} &= \iint u^{(0)}(r_1, r_2)^* \cdot \frac{e^2}{4\pi\varepsilon_0 r_{12}} \cdot u^{(0)}(r_1, r_2) d\tau_1 d\tau_2 \\ &= \frac{14e^2}{\pi^3 \varepsilon_0 a_0^6} \iint \frac{\exp\left[-\frac{4(r_1+r_2)}{a_0}\right]}{|\vec{r}_1 - \vec{r}_2|} d\tau_1 d\tau_2 \\ &= \frac{5e^2}{16\pi\varepsilon_0 a_0}. \end{aligned} \tag{4.10}$$

Finally, the estimated ground energy of helium atom is approximately,

$$\begin{aligned} E_0 \approx E_0^{(0)} + E_0^{(1)} &= -\frac{4m_e e^2}{(4\pi\varepsilon_0)^2\hbar^2} + \frac{4}{5}\frac{m_e e^2}{(4\pi\varepsilon_0)^2\hbar^2} \\ &= -\frac{11}{4}\frac{m_e e^2}{(4\pi\varepsilon_0)^2\hbar^2} = -74.83 \text{ eV}. \end{aligned} \tag{4.11}$$

While the experimental result is

$$E_0 = -78.98 \text{ eV}. \tag{4.12}$$

Obviously, the relative error of the result of first level perturbation approach to that of experimental data is 5.3%, which reveals the validity of the approach [1–6]. The estimated first ionization energy of helium atom He becoming helium ion He^+ is 20.408 eV with first order approximation and 23.065 eV with variational method, while the experimental data is 24.590 eV [5].

The significant feature of the result here indicates the interactions between two electrons, and the actual ground energy of helium atom doesn't equal to any type of simple superposition of single election but reveals the wholeness of the electronic system.

3. **Hydrogen molecules**

Hydrogen molecule has two electrons moving in the electrostatic field (Coulomb field) of the two nuclei [6, 7].

Since the number of electrons and nuclei of hydrogen molecule is 2, respectively, the problem is more complex. Ignore the interactions between spins of two electrons and between spin of electron and orbit, and also neglect the kinetic energy of nuclei (Born–Oppenheimer approximation), the Hamiltonian of hydrogen molecule is simplified as,

$$\widehat{H}^{(0)} = -\frac{\hbar^2}{2m_e}(\nabla_1^2 + \nabla_2^2) - \frac{e^2}{4\pi\varepsilon_0 r_{A1}} - \frac{e^2}{4\pi\varepsilon_0 r_{B1}} - \frac{e^2}{4\pi\varepsilon_0 r_{A2}} - \frac{e^2}{4\pi\varepsilon_0 r_{B2}} + \frac{e^2}{4\pi\varepsilon_0 r_{12}} + \frac{e^2}{4\pi\varepsilon_0 R_{AB}}. \tag{4.13}$$

In Eq. (4.13), r_{12} is the distance between two electrons, r_{A1} expresses distance of nucleus A to electron number 1, ∇_1^2 and ∇_2^2 express nabla symbols of the two electrons' positions of r_1 and r_2, R_{AB} is the distance between two nuclei. The precise analytical solution of hydrogen molecule is also impossible, so it could obtain an approximate solution as well.

The electrostatic interaction between two electrons can also be seen as perturbation term, and the wave functions of two electrons in ground state are used to set up the approximate wave function, the problem can be solved approximately. The approximate values from Heitler, London and Sugiura for binding energy of two atoms is 3.20 eV with bond length of 0.80 A, while approximate values from Wang for binding energy of two atoms is 3.782 eV with bond length of 0.75 A. The experimental data of binding energy of hydrogen molecule is 4.48 eV [6, 7]. The ionization energy of hydrogen molecule H_2 becoming ion of hydrogen molecule H_2^+ is 15.38 eV [5]. Analogously, the actual ground energy of hydrogen molecule doesn't equal to any type of superposition of single hydrogen atom.

4. **Essence of differences**

The differences of states and ground energies of hydrogen atom, helium atom and hydrogen molecule are clear, which reveal the complex interactions among electrons, the states and energy of atom or molecule can only be obtained by solving the specific problem itself as a whole completely.

4.3 Electronic Gas in Metals, First Ionization Energy of Atoms and Work Function of Bulk Metals

As a rough approximation, the electrons in metals can be considered as electronic gas that moves freely [5, 8]. For a cubic block of metal, its electronic energy level is

Table 4.1 Comparison of first ionization energy of metallic atoms and their work function of bulk metals

Metal	W	Ni	Ta	Ag	Cs	Pt	Cr
ϕ/eV [8]	4.5	4.6	4.2	4.8	1.8	5.2	4.6
I/eV[a] [9]	9.2	8.0	7.0	7.8	3.5	9.4	6.8

[a] Data is estimated from diagram of literature [9]

$$E_n = \frac{\pi^2\hbar^2}{2m_eL^2}(n_x^2 + n_y^2 + n_z^2), \quad n_x, n_y, n_z = 1, 2, 3, \ldots . \tag{4.14}$$

Let $n^2 = n_x^2 + n_y^2 + n_z^2$, then Eq. (4.14) reads,

$$E_n = \frac{\pi^2\hbar^2}{2m_eL^2}(n^2). \tag{4.15}$$

Obviously, the energy gap of levels ΔE_n now becomes quite small. The number of electrons dN that can be accommodated in the range of $(n, n + dn)$ is [5],

$$dN = \pi n^2 dn. \tag{4.16}$$

Furthermore, it derives [5],

$$E_f = \frac{\hbar^2}{2m_e}(3\pi^2 N/\Omega)^{2/3}. \tag{4.17}$$

In Eq. (4.17), E_f is Fermi energy level, N is total number of free electrons, and $\Omega = L^3$ represents the volume of the cubic block of metal.

In general, the free electrons are bounded within the bulk of metals. If sufficient energy ($\geq$ work function ϕ) is given, then a free electron can escape from the bulk of metal.

Table 4.1 shows comparison of some data of first ionization energy I of metallic atoms and work function ϕ of bulk metals, which reflect the group interaction and effect of electrons to the energy to drag an electron from atom and the corresponding metallic bulk [8].

4.4 Electrical Resistivity of Pure Metal

The valence electrons of a metal are relatively free and can move in the whole metal. When no external electric field is applied, there is no directional electric current in metal. But as an external electric field is applied [10], the directional movement of electrons is formed in the metal, that is, the electric current is formed. Moreover, when the applied electric field is constant, the obtained electric current is also constant and will not increase with the sustaining of the applied electric field. This shows that the acceleration of the electric field in the metal is "hindered" or "balanced" by some kind of retardation. This obstruction is called electrical resistance. The electrical resistance is the obstacle to the movement of electrons caused by the thermal vibration of atoms at their equilibrium positions due to the action of heat of temperature and the lattice distortions due to the action of temperature and lattice defects. When electrons are accelerated by electric field, they will "collide" with the distortions of lattice and therefore lose energy. The distance between two collisions is the average free path $\overline{L}$, and if the average elapsed time is $\overline{\tau}$, then the average speed is $\overline{u} = \overline{L}/\overline{\tau}$.

Conductivity σ is the reciprocal of resistivity ρ, $\sigma = 1/\rho$. The relationship between conductivity σ, $\overline{L}$, charge e, electron mass m and effective electron number n_e in electron motion is as follows [10],

$$\sigma = \frac{1}{\rho} = \frac{e^2}{2m} n_e \overline{\tau} = \frac{e^2}{2m} n_e \frac{\overline{L}}{\overline{u}}. \tag{4.18}$$

Since the electrical resistance is caused by both the thermal vibration of atoms at their equilibrium positions due to the heat action of temperature and the lattice distortions, the electrical resistance can be further written as,

$$\rho = \rho_T + \rho_n. \tag{4.19}$$

In Eq. (4.19), ρ_T is caused by thermal vibration induced temperature changes, ρ_n is the term caused by lattice distortions and defects.

At higher temperature, with the increase of temperature from absolute zero degree, atom (lattice) deviates from its equilibrium position due to thermal vibration. The mean square value of atomic amplitude increases linearly with temperature, and the average free path of electrons is inversely proportional to the mean square value of atomic amplitude, so the relationship between the electrical resistivity of pure metal and temperature is,

$$\rho_t = \rho_0(1 + \alpha T). \tag{4.20}$$

In Eq. (4.20), α is the coefficient of electrical resistivity with respect to absolute temperature, for pure metals, the value of α is about 4×10^{-3}, for transition metals, especially ferromagnetic metals, α is about 1×10^{-2}.

At lower temperature, $T \ll \theta_D$, electrical resistivity is proportional to T^5, here $\theta_D = h\nu_m/k$, θ_D is called Debye temperature, ν_m is the maximum frequency of atomic vibration. Table 4.2 shows Debye temperature of some elements.

Obviously, the appearance of electrical resistance is caused by the group interactions of movement of electrons and the thermal vibration of atoms at their equilibrium positions due to the action of temperature and the lattice distortions due to the action of temperature and lattice defects.

Table 4.2 Debye temperature of some elements

Element	Na	Ti	Al	Mn	Diamond
θ_D (K)	158	420	428	450	2230

4.5 Summary

The magnitudes of energies to drag an electron from hydrogen atom, helium atom, hydrogen molecule, and bulk metals are quite different, which reflects the group effect and interactions of electrons in the corresponding system indeed; the electrical resistivity of electrons in metals reveals the interactions between electrons and lattices; appropriate models can describe the interactions in specific cases of the corresponding system approximately.

References

1. B. Wu, *Quantum Mechanics* (Springer, Singapore, 2023)
2. T.K. Das, *Quantum Mechanics, Axiomatic Approach and Understanding Through Mathematics* (Springer, Singapore, 2023)
3. P. Kok, *A First Introduction to Quantum Physics* (Springer, Cham, 2023)
4. J.-L. Basdevant, *Lectures on Quantum Physics with Problems*, Exercises and Solution (Springer, Cham, 2023)
5. J.Y. Zeng, *Quantum Mechanics*, vol. 1, 5th edn. (Science Press, Beijing, 2013)
6. S.X. Zhou, H. Chen, *A Course in Quantum Mechanics*, 2nd edn. (Higher Education Press, Beijing, 2009)
7. Y. Shu, X. Chen, S. Duan, R. Liao, How to determine the equilibrium bond distance of homonuclear diatomic molecules: a case study of H_2. Univ. Chem. **39**(7), 1–8 (2024)
8. J.X. Fang, D. Lu, *Solid State Physics*, vol. 1. (Shanghai Science & Technology Press, Shanghai, 1984)
9. S.L. Chu, *Atom Physics*, 2nd edn. (Higher Education Press, Beijing, 2022)
10. R. Wang, *Physical Properties of Metallic Materials* (Metallurgical Industry Press, Beijing, 1993)

Chapter 5
Ising Model in Phase Transition and Elementary Excitation

Abstract Ising model in phase transition, elementary excitation and percolation approach used in solid, order–disorder transition in multi-component alloys, cooperative effects and collective phenomena, and van der Waals' gas are presented in this chapter briefly, which reveals the group actions of the system completely. Besides, Landau's phase transition theory regulates phase transition in term of breaking of symmetry and variation of order parameters; renormalization group considers the scaling invariance of a system with infinite correlation length.

Keywords Phase transition · Ising model · Elementary excitation · Percolation · Order–disorder transition · Cooperative effect · Self-organizing · Van der Waals gas · Renormalization · Collective phenomenon · Order parameter

5.1 Introduction

A lot of cooperative effects and collective phenomena appear in nature, such as phase transition and elementary excitation in solid, which have been described by using Ising model and Magnonics, respectively, to reveal the synergetic interactions among the "elementary site" in the system. The wholeness appearances of the system result from the group actions completely.

Mean field theory is a kind of phenomenological approach in earlier time, such as von de Waals equation, Ising model, etc., which gained great successes in understanding and prediction of relevant phenomena [1–3]. The common feature of different models is the "order-parameter", which is with different actual meaning for different system, but it is always connected with a so called "mean field". By using this "mean field" the actual interaction within the elementary components is depicted abstractly. In this chapter, Ising model in phase transition, elementary excitation in magnetic system, percolation model and Landau theory in complex "ordered-disordered transition in multi-component alloy system", van der Waals' gas, and renormalization group are demonstrated concisely.

M. Zheng and J. Yu, *Systems Theory for Engineering Practice*,
https://doi.org/10.1007/978-981-97-9342-6_5

5.2 Ising Model in Phase Transition

Ising model in phase transition is a very usual model in nowadays [1–3].

There exist other magnetic lattice models beside Ising model (Lenz 1920; Ising 1925), such as Heisenberg model (Heisenberg 1928), X–Y model (Matsubara and Matsuda 1956), O(n)-vector model, spherical model (Berlin and Kac 1952), Potts model (Potts 1952), etc., but Ising model is relatively simple.

In the Ising model, a spin is projected only on direction of a magnetic field h (magnetic phase transition as an example), and there are only two values to be considered, i.e., $\sigma = +1$ and $\sigma = -1$, in general.

In addition, Ising model has characteristic of the discrete quantum value of its energy spectrum due to its discrete spin projection, $\sigma = +1$ or $\sigma = -1$. On the other hand, spin projections considered in direction of the magnetic field, which makes the complication of the problems simpler. Therefore, Ising model is the most simple and convenient mode to demonstrate phase transitions phenomena easily.

In Ising model, spins σ_i means a spin projection on i-th lattice sites, which can be only projected on direction of the magnetic field h. These projections can be equal to values of $+1$ or -1 only.

The value of $\sigma_i = +1$ represents that the projection of spin on i-th site is along the magnetic field h, and it will be denoted by a symbol "↑". In the opposite case, $\sigma_i = -1$ reveals that the projection of spin on i-th site is against the magnetic field h, and it will be denoted by a symbol "↓".

The energy of spin σ_i in the magnetic field h is $-\mu h\sigma_i$, where μ is the value of the spin's magnetic moment. Each spin could only have two possible values of energy in magnetic field, i.e., $-\mu h$ and $+\mu h$, which indicates that the spin-magnetic field interaction is a dual-level system. The exchange interaction energy of spin pairs is $-\sum_{<i,j>} \varepsilon(\vec{r}_i, \vec{r}_j) \cdot \sigma_i\sigma_j$, in which the sum goes over all pairs $<i, j>$ of spins on the lattice. Positive sign of $\varepsilon(\vec{r}_i, \vec{r}_j)$ is corresponding to the ferromagnetic case when two spins have the same projected value and contribute lower interaction energy.

In nearest neighbor approximation, it neglects interactions of spins over longer distances except the nearest neighbors (e.g., a lattice constant a), the exchange interaction energy of spin pairs reduces as $-\varepsilon \sum_{<i,j>n.n.} \sigma_i\sigma_j$, here "*n.n.*" in the sum means that the sum only goes over those pairs of a spin's "nearest neighbors". This kind of the Ising model is named the *n.n.* Ising model.

Furthermore, the Hamiltonian of the system is defined as [1–3],

$$H = -\varepsilon \sum_{<i,j>n.n.} \sigma_i\sigma_j - \mu h \sum_{i=1}^{N} \sigma_i. \tag{5.1}$$

In the following we only employ *n.n.* Ising model, so the notation "*n.n.*" in the sum is omitted for simplicity. Thus, the Hamiltonian of the system is written as,

$$H = -\varepsilon \sum_{<i,j>} \sigma_i \sigma_j - \mu h \sum_{i=1}^{N} \sigma_i. \tag{5.2}$$

If the number of nearest neighbors (the so called coordinate number) of site is γ, then the total number of sum over those pairs of nearest neighbors of all sites is $\gamma N/2$. Assume that the total number of spin in upper orientation (↑) is N_+, the total number of spin in lower orientation (↓) is N_-, the total number pairs of the neighbor spins being upper (↑↑) is N_{++}, the total number pairs of the neighbor spins being lower (↓↓) is N_{--}, and the total number pairs of the neighbor spins being in opposite orientations (↑↓) is N_{+-}. Then it derives following relationships [1–3],

$$N = N_+ + N_-, \tag{5.3}$$

$$\gamma N_+ = 2N_{++} + N_{+-} \rightarrow N_{+-} = \gamma N_+ - 2N_{++}, \tag{5.4}$$

$$\gamma N_- = 2N_{--} + N_{+-} \rightarrow N_{--} = \frac{1}{2}\gamma N + N_{++} - \gamma N_+. \tag{5.5}$$

Thus the Hamiltonian of the system is rewritten as,

$$H = -4\varepsilon N_{++} + 2(\gamma\varepsilon - \mu h)N_+ - \left(\frac{1}{2}\gamma\varepsilon - \mu h\right)N. \tag{5.6}$$

The partition function is [1–3],

$$\begin{aligned} Q(\beta, h) &= \sum_{S_I} \cdots \sum_{S_i} e^{-\beta E_I\{S_i\}} \\ &= e^{N\beta\left(\frac{1}{2}\gamma\varepsilon - \mu h\right)} \sum_{N_+=0}^{N} e^{-2\beta(\gamma\phi - \mu h)N_+} \sum_{N_{++}} g(N_+, N_{++}) e^{4\beta\varepsilon N_{++}}. \end{aligned} \tag{5.7}$$

The subscript "I" in Eq. (5.7) means Ising model. The second summation in Eq. (5.7) can be done for determined N_+; $g(N_+, N_{++})$ expresses the total number of configuration of the given N_+ and N_{++}, which is a complex function of N_+ and N_{++}.

Actually, the determinations of N_+ and N_{++} are difficult except for 1 and 2 dimensions, it has to get an approximation for higher dimensions [1–3].

In general, Ising model can be solved by using mean field approach, such as Bragg-Williams approximation.

In Bragg-Williams approximation, it assumes,

$$N_{++} \approx \frac{1}{2}\gamma N \left(\frac{N_+}{N}\right)^2. \tag{5.8}$$

Furthermore, a long range order parameter L is introduced, such that,

$$\frac{N_+}{N} = \frac{1}{2}(1+L), \quad \frac{N_-}{N} = \frac{1}{2}(1-L), \quad L = (N_+ - N_-)/N, \quad (-1 \le L \le 1). \tag{5.9}$$

Then Bragg-Williams approximation gives,

$$\begin{aligned} H &= -\frac{1}{2}\gamma\varepsilon N(1+L)^2 + (\gamma\varepsilon - \mu h)N(1+L) - \left(\frac{1}{2}\gamma\varepsilon - \mu h\right)N \\ &= \left(-\frac{1}{2}\gamma\varepsilon L^2 - \mu h L\right)N. \end{aligned} \tag{5.10}$$

$$Q(\beta, h) = \sum_{S_i} \exp\left[\beta N\left(\frac{1}{2}\gamma\varepsilon L^2 + \mu h L\right)\right]. \tag{5.11}$$

For given L or N_+, the configuration number is

$$C_{N_+}^N = \frac{N!}{N_+!(N-N_+)!} = \frac{N!}{(\frac{1}{2}N(1+L))!(\frac{1}{2}N(1-L))!}. \tag{5.12}$$

Thus,

$$\begin{aligned} Q(\beta, h) &= \sum_{N_+=0}^{\infty} \frac{N!}{N_+!(N-N_+)!} \exp\left[\beta N\left(\frac{1}{2}\gamma\varepsilon L^2 + \mu h L\right)\right] \\ &= \sum_{L=-1}^{1} \frac{N!}{(\frac{1}{2}N(1+L))!(\frac{1}{2}N(1-L))!} \exp\left[\beta N\left(\frac{1}{2}\gamma\varepsilon L^2 + \mu h L\right)\right]. \end{aligned} \tag{5.13}$$

As to a macroscopic system, $N \gg 1$ and $N_+ \gg 1$, only keep the 1st term of summation (Stirling approximation), $\ln(N!) \approx N \ln N - N$, and in this case denote L by $\overline{L}$ now, then it gives,

$$\frac{1}{N}\ln Q \approx \beta\left(\frac{1}{2}\gamma\varepsilon \overline{L}^2 + \mu h \overline{L}\right) - \frac{1+\overline{L}}{2}\ln\left(\frac{1+\overline{L}}{2}\right) - \frac{1-\overline{L}}{2}\ln\left(\frac{1-\overline{L}}{2}\right). \tag{5.14}$$

The maximum of Q with respect to $\overline{L}$, i.e. $\partial Q/\partial \overline{L} = 0$, gives,

$$\frac{1}{2}\ln\left(\frac{1+\overline{L}}{1-\overline{L}}\right) = \frac{\mu h}{k_B T} + \frac{\gamma\varepsilon \overline{L}}{k_B T}, \quad \text{or } \overline{L} = \tanh\left(\frac{\mu h}{k_B T} + \frac{\gamma\varepsilon \overline{L}}{k_B T}\right), \quad \beta \equiv 1/(k_B T). \tag{5.15}$$

Simultaneously free energy of the system is

$$F(T, h) = -k_B T \ln Q = \frac{N}{2}\gamma\varepsilon\overline{L}^2 + \frac{1}{2}k_B T \ln\left(\frac{1-\overline{L}^2}{4}\right). \tag{5.16}$$

In case of $\varepsilon > 0$ and $h = 0$, it gives the *phase transition temperature* (or critical temperature of phase transition) T_c,

$$T_c = \gamma\varepsilon / k_B. \tag{5.17}$$

At the same time the value of *long range order parameter* $\overline{L}$ reads,

$$\overline{L} = \tanh\left(\frac{\gamma\varepsilon\overline{L}}{k_B T}\right) = \tanh\left(\frac{T_c}{T}\overline{L}\right). \tag{5.18}$$

Equation (5.18) indicates that $\overline{L} = 0$ for $T > T_c$ and $\overline{L} \neq 0$ for $T < T_c$, which means a *spontaneous magnetization* in condition of $T < T_c$.

The *phase transition temperature is surely determined by the exchange interaction energy* ε *among elementary lattices or sites*, which reveals the *collective and synergetic actions of group in the system.*

5.3 Elementary Excitation in Solid

As the distance between spins is small enough so that the wave functions of the two neighbor particles overlap, spins could couple each other through a quantum mechanical mechanism [4–6], which is called the exchange interaction. The simplest treatment for this interaction is the nearest-neighbor approximation, which gives the Heisenberg Hamiltonian and the corresponding spin wave.

Considering a linear chain of classical spins illustrated in Fig. 5.1. Assume that the chain has N spins with magnitude S, which are coupled with nearest-neighbor exchange interaction, uniformly separated by a distance a. Preliminarily, it considers the Heisenberg interaction, then the total exchange energy is given by

$$U_{exc} = -2J\sum_i \vec{S}_i \cdot \vec{S}_{i+1}. \tag{5.19}$$

In Eq. (5.19), J indicates the nearest neighbor exchange parameter and $\vec{S}_i$ expresses the classical spin at i-th site. In the ground state, all spins are in the same orientation (parallel), see Fig. 5.1a, therefore, $\vec{S}_i \cdot \vec{S}_{i+1} = S^2$, the total exchange energy of the system is $U_{exc} = -2JNS^2$.

If there is an effective field (induction) $\vec{B}_T$ representing all interactions on the spin $\vec{S}_i$ of i-th site, then the energy of the moment has the form of $U_i = -\vec{\mu}_i \cdot \vec{B}_T \equiv g\mu_B\vec{S}_i \cdot \vec{B}_T$. In linear spin chain there is an effective field, which is induced by the

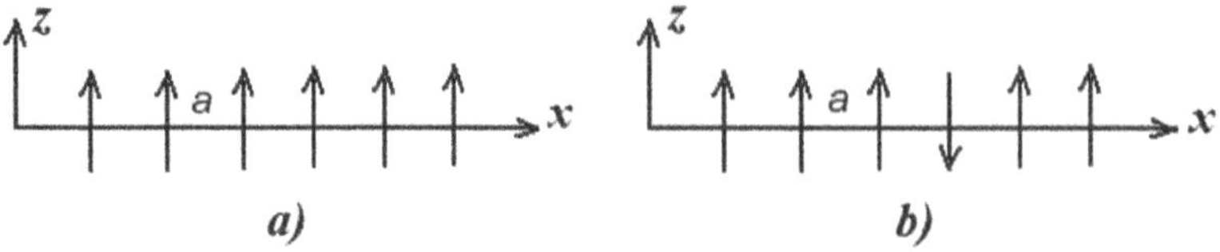

Fig. 5.1 Linear chain of classical spins. **a** Linear chain of classical spins with all spins in same orientation (ground state); **b** linear chain of spins with a spin opposite orientation (excited state)

exchange interaction,

$$\vec{H}_{exc}^{eff} = -\frac{2J}{g\mu_B\mu_0}(\vec{S}_{i-1} + \vec{S}_{i+1}). \tag{5.20}$$

So, besides the exchange interactions with their neighbors, the spins are subject to an applied static magnetic field $\vec{H}$, therefore the total field acting on the spin $\vec{S}_i$ is $\vec{B}_T = \mu_0(\vec{H} + \vec{H}_{eec}^{eff})$. The torque acting on the moment has the form of $\vec{\tau} = \vec{\mu}_i \times \vec{B}_T$, using the torque equation, $d(\hbar\vec{S})/dt = \vec{\tau}$, it derives the motion equation of spin $\vec{S}_i$,

$$d\vec{S}/dt = -\gamma\mu_0\vec{S}_i \times (\vec{H} + \vec{H}_{exc}^{eff}) \text{ with } \gamma = g\mu_B/\hbar. \tag{5.21}$$

Equation (5.21) reveals that the motion of the spin $\vec{S}_i$ in i-th site is coupled to the motions of its neighboring spins, which indicates a *collective excitation of this kind of motion.* Further study indicates that the collective excitation of this kind of motion behaves as a wave with the frequency of $\omega_k = \gamma\mu_0 H + \frac{4JS}{\hbar}[1 - \cos(ka)]$. This is the so called dispersion relation that reflects the dependence of spin's wave frequency on the wave number.

5.4 Site-Bond Percolation and Phase Transition in Diluted Magnetic System

Percolation model is a very effective approach to deal with phase transition problems in disordered system. There are typically three kinds of percolation processes, i.e., bond percolation, site percolation, and site-bond mixture percolation [7, 8]. The basic problem in site-bond mixture percolation remains unclear though the fundamental ones in both bond percolation and site percolation have been clarified [7, 8], two approaches, i.e., renormalization and numerical simulation, were employed to conduct the relevant study frequently, valuable results are obtained. While on the other hand, an alternative phenomenological approach was proposed to depict the processes in site-bond mixture percolation forthrightly [8].

Phase transition in diluted magnetic system is a typical site-bond mixture percolation, which attracts interesting of both statistics and magnetics [8]. Here in this section, it is concisely described by using the phenomenological approach with ingenuity analytically.

In pure bond percolation, all rigid sites are occupied by "magnetic atom" for example [8]. As the bond probability P_b is higher than its critical value or threshold, P_{bc}, an infinite percolation cluster builds, which corresponds to a magnetic status or "conductive status". The critical value, P_{bc}, is dependent on its microstructure and topological dimension [7, 8]. The *bond here plays the role of short-range-order*.

In condition of the coordinate number (or number of the nearest atoms) Z_{e0} given, there exists following empirical relationship [7],

$$P_{bc} \cdot Z_{e0} = C. \tag{5.22}$$

In Eq. (5.22), C is a constant, which reflects the characteristic of topological dimension, $C = 2$ for 2-d and $C = 1.45$ for 3-d problems [7]. Equation (5.22) reveals that the value of P_{bc} increases with decrease of Z_{e0} in the same topological dimension so as to build an infinite percolation cluster (long-range-order) in the system.

On the other hand, in pure site percolation, all bonds are closed to provide a local "passage", while sites are randomly occupied, if the occupied site probability P_s is higher than its critical value or threshold, P_{sc}, an infinite percolation cluster builds analogically, which corresponds to a "conductive status", too [7, 8]. There critical value, P_{sc}, is dependent on its microstructure and topological dimension as well [7, 8]. In condition of the coordinate number (or number of the nearest atoms) Z_{e0} given, there exists following empirical relationship [8],

$$(\alpha P_{sc} + \beta) \cdot Z_{e0} = C. \tag{5.23}$$

In Eq. (5.23), the constant C is the same as those in Eq. (5.22) [8], while α and β in Eq. (5.23) are new constants, $\alpha = 1.65$ and $\beta = -0.50$ for 2-d and $\alpha = 1.07$ and $\beta = -0.09$ for 3-d problems, respectively [8]. Equation (5.23) also reflects that the value of P_{sc} increases with decrease of Z_{e0} in the same topological dimension so that the infinite percolation cluster could be built in the system.

However, in site-bond mixture percolation [7, 8], both sites and bonds are not all filled, but partially occupied with certain probabilities. As to a fixed microstructure, if its occupied probability of sites is P_s, only when the closed bond probability P_b is greater than a instant critical value, says, P'_{bc}, an finite percolation cluster could form in the system. It is sure that the value of $P_{bc}{}'$ depends on P_s for a system with fixed mcrostructure, which can be called a phase boundary relationship of site-bond mixture percolation process, i.e.,

$$P'_{bc} = F(P_s). \tag{5.24}$$

The phase boundary relationship of site-bond mixture percolation process reveals the cooperation between "site" and "bond" in forming the infinite percolation cluster coordinately.

Furthermore, assume an alternative function $f(P_s) = C/F(P_s)$, it yields an equivalent form of Eq. (5.24),

$$P'_{bc} \cdot f(P_s) = C. \tag{5.25}$$

Obviously, Eq. (5.25) can be seen as the generalized form of both Eq. (5.22) and (5.23). In other words, either Eq. (5.22) or (5.23) is the special case of Eq. (5.25) at $P_s = 1$, and $P'_{bc} = P_c$, or $P_s = P_{sc}$, and $P'_{bc} = 1$, respectively.

In general, mathematically the function $f(P_s)$ can be expanded as a series of P_s, i.e.,

$$f(P_s) = \sum_{k=-\infty}^{\infty} a_n \cdot P_s^n. \tag{5.26}$$

Keep in mind that both pure bond percolation and pure site percolation are all special cases of site-bond percolation process, the expression Eq. (5.26) could be simplified. Finally, a phase boundary relationship of site-bond mixture percolation process is put forward [8], i.e.,

$$P'_{bc} \cdot (a_0 + a_1 \cdot P_s) = C, \tag{5.27}$$

In Eq. (5.27), the coefficients a_0 and a_1 take following value,

$$a_0 = C - P_{sc} \cdot a_1, \quad a_1 = C \cdot (1 - P_{bc})/[P_{bc} \cdot (1 - P_{sc})]. \tag{5.28}$$

Table 5.1 shows estimated values of a_0 and a_1 for some typical microstructures [8].

Subsequently, in a diluted magnetic system, says, $A_{P_s}B_{1-P_s}$, magnetic element A occupies on the sites with probability of P_s, while the coupling probability of a magnetic element A with its nearby magnetic elements is bond probability P_b, then

Table 5.1 Estimated values of a_0 and a_1 for some typical microstructures

Dimension	Structure	P_{bc}	P_{sc}	a_1	a_0	C
2-d	S.Q.	0.500	0.593	5.000	– 1.000	2
	Tri.	0.347	0.500	7.517	– 1.759	
	H.G.	0.653	0.698	3.524	– 0.460	
	Kagome	0.524	0.653	5.236	– 1.419	
3-d	F.C.C.	0.119	0.198	13.385	– 1.200	1.45
	B.C.C.	0.179	0.245	8.809	– 0.708	
	S.C.	0.247	0.311	6.416	– 0.545	
	Diamond	0.388	0.428	3.998	– 0.261	

the general phase boundary relationship, Eq. (5.27), could be used to deal with such site-bond mixture percolation problem in principle.

Furthermore, according to Potts model (Ising model as well), the coupling probability between magnetic elements at absolute temperature T is [8],

$$P_b = 1 - \exp[-J/(k_b \cdot T)]. \tag{5.29}$$

In Eq. (5.29), k_b is Boltzmann constant, J is the coupling constant of magnetic elements.

Moreover, if the Curie temperature of the "pure magnetic system" is denoted by T_{c0}, it yields following expression for P_{bc},

$$P_{bc} = 1 - \exp[-J/(k_b \cdot T_{c0})]. \tag{5.30}$$

Finally, the phase transition process of a diluted magnetic system $A_{P_s}B_{1-P_s}$, can be described by a site-bond mixture percolation one. The phase transition temperature T_c' of a diluted magnetic system $A_{P_s}B_{1-P_s}$ can be obtained [8],

$$\frac{T_c'}{T_{c0}} = \left[\ln\left(\frac{a_1 + a_0 - C}{a_1 + a_0}\right)\right] \Big/ \left[\ln\left(\frac{a_1 \cdot P_s + a_0 - C}{a_1 \cdot P_s + a_0}\right)\right]. \tag{5.31}$$

Equation (5.31) is phase boundary relationship of diluted magnetic system $A_{P_s}B_{1-P_s}$ under phenomenological approach with Potts model approximation.

Equation (5.31) was used to estimate phase transition of a series of diluted magnetic systems, satisfied results are obtained, Fig. 5.2 cites one case [8].

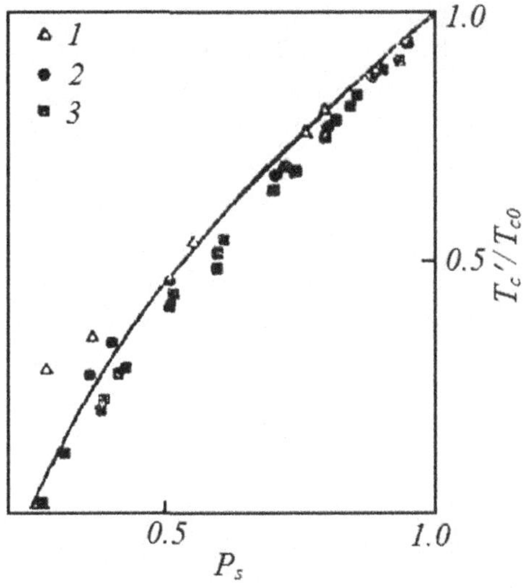

Fig. 5.2 Phase transition of diluted magnetic systems with B.C.C. structure. Δ denotes FeF_2: Zn system; • denotes $CoCsCl_5$: Zn system; ■ denotes MnF_2: Zn system

5.5 Order–Disorder Transition of Binary Ordered Alloys

Order–disorder transition of binary ordered alloys is called super-lattice phase transition as well, which is a very important issue for high strength material research. The brittleness is its fatal disadvantage in use of ordered alloy, the appropriate decrease of orderliness could be an effective approach to enhance its toughness by non stoichiometry or third element addition [9, 10].

The percolation mechanism of order–disorder transition of binary ordered alloys and ternary doping was studied in literatures [9, 10]. Here in this section it is outlined.

As to body centered cubic (*B.C.C.*) and face centered cubic (*F.C.C.*) structural types of ordered alloys, their general expression can be written as $A_xB_{1\text{-}x}$, in which A and B indicate the two types of atoms, x represents the content in atomic percentage, respectively.

Let ε_{AA}, ε_{BB} and ε_{AB} express the interaction energies between the nearest couples of *AA*, *BB*, and *AB*, respectively; and N_{AA}, N_{BB} and N_{AB} represent the numbers of the nearest couples of *AA*, *BB*, and *AB* individually, then the Hamiltonian of the system is rewritten as [9, 10],

$$H = N_{AA}\varepsilon_{AA} + N_{BB}\varepsilon_{BB} + N_{AB}\varepsilon_{AB}. \tag{5.32}$$

Furthermore, let N_A and N_B represent the total numbers of atom A and atom B, and z be the coordinate number of the crystal structure of the alloy, individually, then it leads to,

$$\begin{cases} 2N_{AA} + N_{AB} = zN_A \\ 2N_{BB} + N_{AB} = zN_B \end{cases} \rightarrow \begin{cases} N_{AA} = (zN_A - N_{AB})/2 \\ N_{BB} = (zN_B - N_{AB})/2 \end{cases} \tag{5.33}$$

Thus Eq. (5.32) becomes,

$$\begin{aligned} H &= \frac{1}{2}(zN_A - N_{AB})\varepsilon_{AA} + \frac{1}{2}(zN_B - N_{AB})\varepsilon_{BB} + N_{AB}\varepsilon_{AB} \\ &= \frac{1}{2}zN_A\varepsilon_{AA} + \frac{1}{2}zN_B\varepsilon_{BB} + N_{AB}[\varepsilon_{AB} - \frac{1}{2}(\varepsilon_{AA} + \varepsilon_{BB})]. \end{aligned} \tag{5.34}$$

Since the specific form of Eq. (5.34), it could set $\varepsilon_{AA} = \varepsilon_{BB} = -\varepsilon_{AB} = J > 0$, by ignoring the constant terms in Hamiltonian, it yields,

$$H = J\sum_{i<i,j>} s_i \cdot s_j, \quad s_i = \begin{cases} 1, & atom\,A, \\ -1, & atom\,B. \end{cases} \tag{5.35}$$

Recall the general relationship between bond percolation and Ising model, i.e., Eq. (5.29), the critical bond percolation probability $P_b'(x)$ can be estimated for each content percentage x of binary ordered alloy at its order-disorder transition temperature T_c' analogically,

$$P_b'(x) = 1 - \exp[-J/(k_b \cdot T_c')]. \tag{5.36}$$

Since the super-lattice phase transition is an ordering phenomenon in binary alloys, which might have some differences from the general percolation at its critical point perhaps. Monte-Carlo simulations with Metropolis sampling were employed to modeling the order-disorder transition [9]. As to *B.C.C.* typed structure, a lattice of 2×20^3 with periodic boundary condition was used in the simulation, and for *F.C.C.* typed structure, a lattice of 4×14^3 with periodic boundary condition was employed in the modeling.

The jumping point of specific heat is appointed as the order–disorder transition point, a critical value of $k_B T_c' = 4.75J$ was obtained for *B.C.C.* structure at the stoichiometric content of $A_{0.5}B_{0.5}$, while other critical value of $k_B T_c' = 2.95J$ was obtained for *F.C.C.* structure at the stoichiometric content of $A_{0.25}B_{0.75}$, which showed a structural dependence of the order–disorder transition point distinctly [9]. Furthermore, the corresponding values of critical bond percolation probability $P_b'(x)$, are 0.1898, and 0.2875 for *B.C.C.* and *F.C.C.* structures, respectively, which differ from those of the general bond percolations obviously, see Table 5.1.

Furthermore, alloys with non stoichiometric contents for *B.C.C.* and *F.C.C.* structures were simulated [9], a series of values of $k_B T_c'/J$ corresponding to different non stoichiometric contents were obtained, which are cited in Tables 5.2 and 5.3. The contents with asterisk "*" in Tables 5.2 and 5.3 denotes the stoichiometric content of the alloy series.

Both Tables 5.2 and 5.3 show that variations of $k_B T_c'/J$ and $P_b'(x)$ with content x tremendously.

Similar to Eq. (5.27), another approximately invariant formula could be proposed formally,

$$P_{bc}' \cdot (b_0 + b_1 \cdot x) = C'. \tag{5.37}$$

If we take $C' \equiv C = 1.45$ for 3-d problem as well, the regressed values of b_0 and b_1 for *B.C.C.* structure are, 18.529 and – 1.5636, obtained with the linear correlation

Table 5.2 Value of $k_B T_c'/J$ *versus* x (content) for *B.C.C.* structure by using Monte-Carlo simulation

x	0.50*	0.45	0.40	0.35	0.30	0.25
$k_B T_c'/J$	4.75	4.20	3.50	2.9	2.20	1.55
$P_b'(x)$	0.1898	0.2119	0.2485	0.2917	0.3653	0.4754

Table 5.3 Value of $k_B T_c'/J$ *versus* x (content) for *F.C.C.* structure by using Monte-Carlo simulation

x	0.50	0.45	0.40	0.35	0.30	0.25*	0.225	0.20	0.175	0.15
$k_B T_c'/J$	7.55	6.10	5.30	4.90	4.00	2.95	2.80	2.22	1.70	1.25
$P_b'(x)$	0.1205	0.1512	0.1719	0.1846	0.2212	0.2875	0.3003	0.3627	0.4447	0.5507

coefficient of $R = 0.9996$, respectively; while for *F.C.C.* structure the regressed values of b_0 and b_1 are, 24.347 and − 0.9159, gained with the linear correlation coefficient of $R = 0.9953$, individually.

For ternary ordered alloy, analogical study was conducted by approximation of incorporating two similar elements as one type of element [10], in which the similarity of two elements is recognized by comparing the values of the interaction energies, ε_{AB}, ε_{BC} and ε_{AC}, the least one indicates two similar atoms.

Furthermore, according to percolation theory [7], systems with different critical temperatures T_c will have different long-range-order at a given temperature T, the lower the T_c is, the smaller the long-range-order of the system, i.e., the smaller the long-range order parameters.

Usually, the crystals nucleates and grows from many places at the same time during solidification, there are always a large number of grain boundaries in solids, and the properties of grain boundaries often have a great influence on the properties of bulk solids. For metallic ordered compounds, each grain may be an ordered domain, and the grain boundary is the domain boundary in general. If the ordered domains on both sides of the grain boundary (domain boundary) are anti-phase, it is called anti-phase domain, and the corresponding domain boundary is called anti-phase domain boundary. In general, the *anti-phase domain boundary has large interface energy*, so *its stability is poor and inter-granular fracture is easy to occur*. The *stronger the long-range-order of domain*, the *greater the possibility of anti-phase domain boundary*, and *the greater the possibility of inter-granular brittle fracture in ordered alloy*. Element substitution is often carried out according to the principle of similarity. *If the effective component variable of the compound is reduced as a result of element substitution, the critical temperature of the substituted system will be lower than that of the stoichiometric material*, so that at a certain temperature, *the long-range order of domains will be small, the possibility of anti-phase domains and anti-phase domain boundaries will be suppressed a little*, and *the chance of inter-granular brittle fracture will be reduced in the ordered alloy*. This is the percolation mechanism of doping and toughening of metal compounds [7, 9, 10].

5.6 Landau's Phase Transition Theory

In Landau theory, the transition between phases is considered to be a change of phases which are with different symmetries [11, 12].

At high temperatures, the phase is with a higher symmetry in general consideration specifically; while at lower temperature, the corresponding phase of the counterpart is usually with a lower symmetry or even ordered one. The transition to the ordered phase corresponds to a breaking of symmetry in a sense. Besides, since the symmetry breaking during phase transition from higher temperature to lower temperature is not a result of any external field, the transition is thus called with a *spontaneous breaking of symmetry*.

In the ordered phase, there coexists actually several distinct ordered phases with similar structures. The structures can be connected to each other by certain symmetry operation, which comprises a group of symmetry, G. The symmetry operation of the group G could complete the transformation from an ordered state into another, which makes all ordered states in coexistence from one of them. This indicates that the disordered phase has an invariance under the transformations of the group G. In Landau theory, all ordered states are considered to have the same Gibbs free energy.

In order to depict the ordered phase an *order parameter* is introduced reasonably, which is a new thermodynamic variable that takes *nonzero value in the ordered phase and vanishes in the disordered phase.* Furthermore, in order to reflect the possibility of the appearance of distinct ordered states, it assumes that the order parameter could be a vector which has several components and denoted by $\vec{\eta}$.

In disordered phase, the order parameter vanishes, i.e.,$\vec{\eta} = 0$. Thus the thermodynamic state associated with the molar Gibbs free energy is defined by the components η_i of the vector $\vec{\eta}$ in additional to the usual variables, such as temperature, pressure and mole fractions.

In Landau theory, the order–disorder transition is associated with the spontaneous breaking of symmetry without any external field. Thus, the molar Gibbs free energy is written as,

$$\partial g/\partial \eta_i = 0. \tag{5.38}$$

Moreover, since g must be a convex function of η_i, then requirement Eq. (5.38) means that the phase transition between ordered phase and disordered phase is corresponding to the minima of g with respect to the order parameter.

Furthermore, in Landau theory for the order–disorder transition, it assumes that $g(\vec{\eta})$ is a polynomial in η_i. In general, it supposes that the polynomial is of fourth degree [11, 12], i.e.,

$$g(\vec{\eta}) = a_0 + a_2 I_2(\vec{\eta}) + a_3 I_3(\vec{\eta}) + a_4 I_4(\vec{\eta}). \tag{5.39}$$

In Eq. (5.39), $I_n(\vec{\eta})$ is a monomial or a sum of monomials of degree n in η_i and the coefficients a_n are dependent of other thermodynamic variables with the exception of η_i [11, 12].

A simplest case in Landau theory, such as the a binary alloy A-B, is that g is a polynomial of fourth degree in a single component η, i.e.,

$$g(\eta) = a_0 + a_2\eta^2 + a_4\eta^4. \tag{5.40}$$

Furthermore, it assumes that the coefficient a_4 is considered to be positive in Eq. (5.40), and the coefficient a_2 is positive in the disordered phase and negative in the ordered phase, i.e.,

$$a_2 = A(T - T_c). \tag{5.41}$$

In Eq. (5.41), $A > 0$.

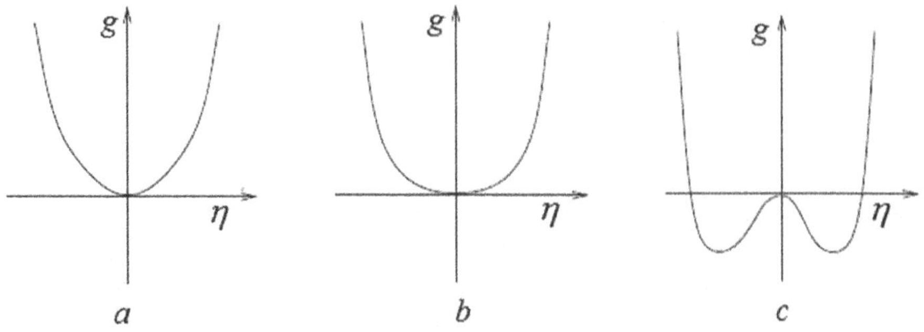

Fig. 5.3 Molar Gibbs free energy g versus order parameter η. **a** The state of temperature above T_c; **b** the state of temperature at T_c; **c** the state of temperature below T_c

The results of the result (5.41) are illustrated in Fig. 5.3. For temperatures above T_c, the function $g(\eta)$ is convex and has a single minimum at $\eta = 0$. For temperatures below T_c, the function $g(\eta)$ is non convex and has twin symmetric minima. In this case we consider the convex hull of $g(\eta)$. The appearance of two minima is a manifestation of spontaneous breaking of symmetry. It is important to note that as the temperature approaches the critical temperature T_c, the minimum continuously approaches zero.

In ordered state, the convex hull has a horizontal line segment which connects the two minima, i.e.,

$$\partial g(\eta)/\partial n = 2a_2\eta + 4a_4\eta^3 = 0. \tag{5.42}$$

It yields,

$$\eta = \pm\sqrt{\frac{|a_2|}{2a_4}}. \tag{5.43}$$

Equation (5.43) is valid for $T < T_c$. For $T > T_c$ it has $\eta = 0$. Therefore, the order parameter vanishes continuously and behaves in lower temperature according to Eq. (5.44),

$$\eta \sim |T_c - T|^{1/2}. \tag{5.44}$$

5.7 Other Approaches Beyond Our Descriptions

Synergetics [13–15], dissipative structure [16–18], emergence phenomenon, the three-body problem in astrophysics, and other nonlinear problems, etc., are usually described in other books extensively, they consisted of other part of systems science.

In recent years, dissipative structure theory, synergetics, catastrophe theory, chaos theory, fractal theory and dynamic systems theory were developed, which all contribute to the achievements of self-organization theory [19, 20]. According to the theory of self-organization, the cooperation and competition among the subsystems in a system produce time structure, space structure and functional structure. This theory is expected to build a bridge between inanimate phenomena and living phenomena. Here, time structure refers to the orderly change of some phenomena or physical quantities with time, such as the beating of heart and the change of brain wave image; Spatial structure refers to the spatial distribution of matter, energy or a certain unit, such as the shape of rivers and the pattern of clouds; Functional structure refers to the generation of functions and the relationship between multiple functions, such as metabolism. A system is called a self-organizing system if there is no specific external intervention when it produces time structure, space structure and functional structure under certain conditions. Its temporal structure, spatial structure and functional structure are collectively called self-organizing structure. Self-organization theory is a theory that studies the general law of self-organization structure.

Laser is a typical self-organizing system. The working substance of a laser contains a large number of atoms or molecules. When the pumping power is below the threshold, these particles "go their own way" and produce spontaneous radiation. This spontaneous emission is disordered and chaotic in phase. Or spontaneous emission is a mob of unorganized and undisciplined photons. The radiation changes from disorder to order near the threshold of the pumping power. Above the threshold of the pumping power, laser is generated. Laser is an organized and disciplined army of photons. The temporal and spatial patterns of laser reflect the "group gymnastics" and "chorus" of atoms. The transformation of laser from disorder to order is entirely the result of cooperation and competition among a large number of atoms. Starting from fluctuation (spontaneous emission), laser is generated through regenerative amplification (stimulated amplification), positive feedback and mode selection (the role of resonant cavity). In the competition of different modes (radiation direction and frequency), only one or a few modes gain the dominant position and become the main mode. This mode of laser is called order parameter. It forces a large number of atoms (subsystems) to act collectively at a unified pace, and to some extent loses the "freedom" of spontaneous radiation. The order parameters are generated by subsystems, which in turn dominate and serve the subsystems. This is the basic principle of self-organization. Servo principle of sequence parameters works to each subsystem.

"Universe is in harmony, and everything is born by itself". Self-organization is very common. The application scope of self-organization theory is also very broad. It emphasizes the *integrity, nonlinearity, dynamics and complexity of the system*, which opens up attractive prospects for studying complex phenomena and establishing complexity theory. It is expected to be applied in many fields [20].

So here in this book we only pay attention on the topics of our issues. The topics of our issues form the powerful branch with both solid physical meaning and distinct mathematical algorithm, which will provide useful approaches to deal with practical problem in systems science.

5.8 Van der Waals' State Equation of Gas

J. D. van der Waals, a Dutch physicist, proposed state equation of gas in 1873, which considered the interactions among molecules. The typical form of van der Waals' state equation of gas is,

$$\left(P + N^2 \cdot \frac{a}{V^2}\right) \cdot (V - Nb) = \frac{M}{\mu} RT. \tag{5.45}$$

In Eq. (5.45), P is the total pressure of the gas; V is the total volume of gas; T is the Kelvin temperature of the gas; N is the total number of molecules of the gas; R is a molar gas constant; a is a van der Waals constant, which represents the interaction force between gas molecules; b is another van der Waals constant, which represents the volume occupied by each molecule of gas. This equation can be applied to reflect the real gas situation, and study the difference between the experimental results and theoretical predictions. It can be used to study the physical properties of gas more accurately, such as pressure, density, and thermodynamic properties, etc.

While the state equation of ideal/perfect gas is,

$$P \cdot V = \frac{M}{\mu} RT. \tag{5.46}$$

Obviously, Eq. (5.46) neglected the interaction force between gas molecules and volume occupied by each gas molecule of the gas system. Comparatively, Eq. (5.45) reveals the overall situations of gas with higher accuracy, which is more close to real gas. The interaction force between gas molecules could be reflected by the apparent total pressure P and the Kelvin temperature T of the gas system macroscopically.

5.9 Scale Transformation and Renormalization Group

Kadanoff et al. proposed scaling invariance by considering the characteristic of correlation length with infinite long scale in 1960s [3].

Logically, since the correlation length is infinitely long, it should be infinitely long at any scale [3]. When the scale is changed within a certain range, the images (phenomena) you see should be all the same. Kadanoff understood this image as the invariance of scale transformation or as "self-similarity". Just like the two-dimensional block depicted in Fig. 5.4, each block contains 9 spins group with a spacing a_0 between two nearby spins; several small pieces form a new "lattice", and the nearest neighbor spacing of the new "lattices" is $3a_0$.

The spin of each block is still characterized by Ising spin, that is, the spin of the j-th spin group is represented by $s_j^B = \pm 1$. Among them, the spin value of the involved small block can be obtained in manner of "minority obeying the majority". In this way, a new Ising lattice whose scale is L times (3 times in this case) is obtained. This

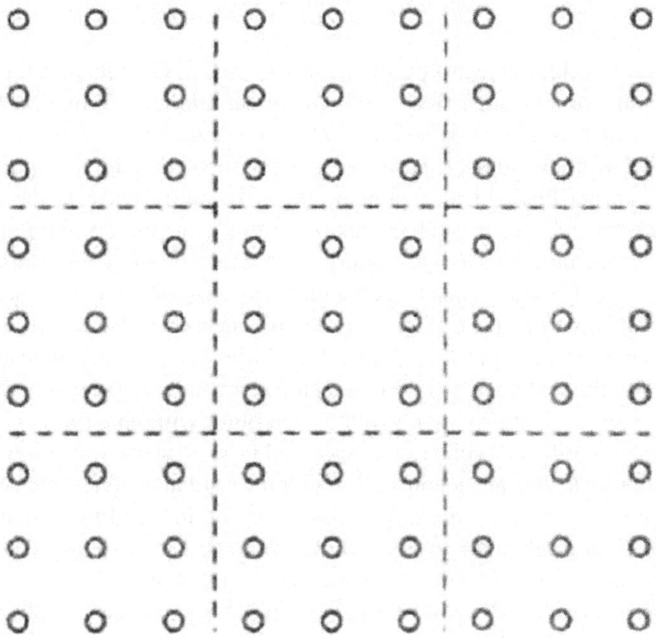

Fig. 5.4 Two-dimensional block with 9 spins group

realizes the scale transformation, also known as Kadanoff transformation. Because the "spin block" gets bigger and bigger, it can also be called "coarse granulation". The Hamiltonian of the system should have the same functional form before and after the scale transformation. This is called the invariance of scale transformation.

Through scaling transformation, the independent parameters (τ, h) describing the system should also be rescaled into (τ_L, h_L), and the following relationship related to the transformation scale L can be obtained,

$$\begin{cases} h_L = hL^x \\ \tau_L = \tau L^y \end{cases}. \tag{5.47}$$

In Eq. (5.47), x and y are power exponents to be determined and take positive values.

If the average Gibbs function of each "small block" is expressed by $g(\tau_L, h_L)$, since each small block contains L^d spins, all of them have the following formula,

$$g(\tau_L, h_L) = L^d g(\tau, h). \tag{5.48}$$

Scaling transformation is equivalent to enlarging the "ruler" to L times, so if one uses it to measure the correlation length, the result should be $1/L$ times, that is,

$$\xi(\tau, h) = L\xi(L^y\tau, L^x h). \tag{5.49}$$

A calculation technique was introduced in 1971 by K. G. Wilson and is known as the "Renormalization Group" (RG) [21]. Under scaling transformation R_s, if there is a point x^* which is an invariance, i.e., $\mu^* = R_s\mu^*$, then x^* is called an invariant. By using this invariant and properly selecting lattice models and approximation methods, it can get the critical exponent, etc. It is surely because the correlation length (the average scale of patch) tends to infinity near the critical point that the system is invariant under this transformation. The invariance under R_s transformation, in terms of physical image, is just like changing a microscope with low magnification to observe the sample, and the difference in details can not be observed from the image near the critical point. The patch at temperature T approaching to critical temperature T_C forms within the system as an interconnected zone of a set of elementary sites with the same phase, which appears embedding each other without definite boundary, as shown in Fig. 5.5, it is the typical feature of 2nd type of phase transition.

The calculation technique not only allows for the critical exponents to be computed but also provides a more profound justification for the scaling and universality assumptions. Renormalization group technique can be used to study such problems for which the physically relevant information is independence of the system's microscopic scales. This generality will be used here to illustrate the renormalization group technique on a physical problem closely related, but not identical, to the calculation of a critical exponent [21].

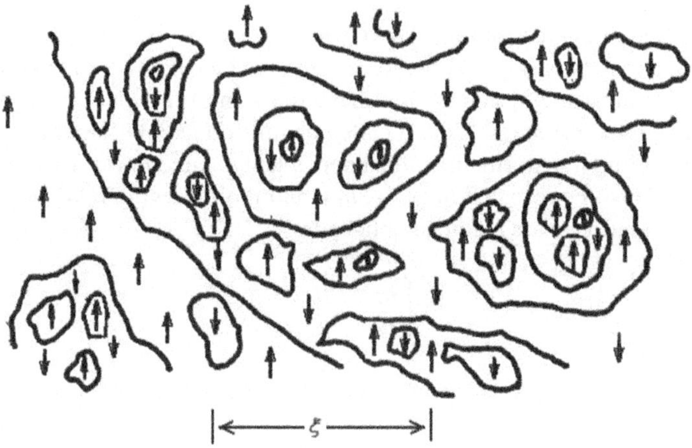

Fig. 5.5 Patch embedding each other

5.10 Concluding Remarks

Ising model in phase transition, elementary excitation and percolation processes in solid, order–disorder transition of multi-component ordered alloys, self-organizing in laser, van der Waals' state equation of gas, and scale transformation and renormalization, all exhibit the cooperative effects and collective phenomena of a system obviously, which reveal the group actions completely. Breaking of symmetry and variation of order parameters of a system lead to phase transition in Landau's theory. The interaction among microscopic elements of a system could be reflected by the apparent phenomena or physical quantities macroscopically. The appearances of wholeness result from the group actions of the systems undoubtedly.

References

1. T. Matsoukas, *Generalized Statistical Thermodynamics, Thermodynamics of Probability Distributions and Stochastic Processes* (Springer, Cham, 2018)
2. O. Civitarese, M. Gadella, *Methods in Statistical Mechanics, A Modern View* (Springer, Cham, 2020)
3. X.X. Liang, *Introduction to Advanced Statistical Mechanics*, 2nd edn. (Science Press, Beijing, 2020)
4. S.M. Rezende, *Fundamentals of Magnonics* (Springer, Cham, 2020)
5. S.O. Demokritov, A.N. Slavin, *Magnonics: Fundamentals to Applications* (Springer, Berlin Heidelberg, 2013)
6. D.D. Stancil, A. Prabhakar, *Spin Waves, Problems and Solutions* (Springer, Cham, 2021)
7. R. Zallen, *The Physics of Amorphous Solids* (A Wiley-Interscience Publication, New York, 1983)
8. M. Zheng, Y. Liu, E. Luo, Site-bond percolation and phase transition in diluted magnetic system. Sci. China (Ser. A) **35**(3), 311–316 (1992)
9. M. Zheng, Y. Liu, Percolation mechanism of super-lattice phase transition of inter-metallic compounds. Acta Physica Sinica **42**(2), 304–308 (1993)
10. M. Zheng, Y. Liu, J. Gao, Z. Jin, Order-disorder transition of ternary ordered alloy. Chin. Phys. Lett. **11**(6), 353–356 (1994)
11. L.D. Landau, E.M. Lifshitz, Statistical physics I, in *Course of Theoretical Physics*, vol. 05, 3rd edn. (Butterworth–Heinenann, Linacre House, Jordan Hill, UK, 2000)
12. M.J. de Oliveira, *Equilibrium Thermodynamics*, 2nd edn. (Springer, Berlin Heidelberg, 2017)
13. H. Haken, *Synergetics, an Introduction* (Springer, Berlin, 1983)
14. H. Haken, *Advanced Synergetics, Instability Hierarchies of Self-organizing System and Devices* (Springer, Berlin, 1983)
15. H. Haken, J. Portugali, *Synergetics Cities: Information, Steady State and Phase Transition* (Springer, Cham, 2021)
16. B. Brogiato, R. Lozano, B. Maschke, O. Egeland, *Dissipative Systems Analysis*, 3rd edn. (Springer, Cham, 2020)
17. H. Mori, Y. Kuramoto, *Dissipative Structures and Chaos*, 3rd edn. (Springer, Berlin, 1998)
18. F.H. Busse, L. Kramer, *Nonlinear Evoluation of Spatio-Temporal Structures in Dissipative Continuous Systems* (Springer, New York, 1990)
19. F. Li, Dynamics of thinking in brain. Science **47** (3), 48–52 (1995) (in Chinese)
20. F. Li, Grand unification of medical sciences. Science **48**(2), 36–40 (1996) (in Chinese)
21. M. Baus, C.F. Tejero, *Equilibrium Statistical Physics, Phases, Phase Transitions, and Topological Phases*, 2nd edn. (Springer, Cham, 2021)

Chapter 6
Brief Description of Probabilistic Multi-objective Optimization of a System

Abstract This chapter gives a brief description of probabilistic multi-objective optimization of a system, which includes the scientific connotation and essence of multi-objective optimization from viewpoint of systems theory. The main thought is to reflect the "simultaneity of multi-objective optimization" by using probabilistic approach from viewpoint of systems theory. The concept of "preferable probability" and its evaluation are overviewed in respects of set theory and probability theory. Besides, robust design is an optimal approach that aims to reduce effect of uncertain factors on product quality, which is a specific case of probabilistic multi-objective optimization and outlined. The fuzzification treatment is utilized to the case of desired value as target in robust design as an auxiliary approach. Additionally, the reliability design and assessment involving optimization with multiple objectives is taken as another specific case of probabilistic multi-objective optimization as well.

Keywords Systems theory · Probability theory · Multi-objective optimization · Simultaneous optimization · Integrity · Preferable probability · Robust design · Reliability design

6.1 Introduction

In a system, optimization is a very usual issue, which means the optimal status of the system as a whole. Many objectives/attributes are usually contained in a system, so the optimization of the system is inevitably an optimization problem with multiple objectives (MOO). The main feature of multi-objective optimization is the "simultaneity of optimization of the multiple objectives" in the system, which means the irreplacement of the multiple objectives in the optimization. However, the previous traditional methods for MOO adopted the "*additive*" *algorithm* together with *normalization/parameterization of the involved attributes for the evaluation indexes*, and some even include *personal factors* [1–5], etc. Actually, "additive" algorithm of the multiple indexes in evaluation has its inherent characteristics, such as replacebility of the involved attributes, non-uniqueness apart from arbitrariness of the normalized

M. Zheng and J. Yu, *Systems Theory for Engineering Practice*,
https://doi.org/10.1007/978-981-97-9342-6_6

scales, etc., which is in fact *not coincident with the real intention of "simultaneous optimization of multiple indexes"*, but *equivalent to taking the form of "union" in the spirits of set theory and probability theory instead* [6]. In fact, in the respect of probability theory, the *essence of "simultaneous optimization of multiple indexes" has the meaning of "joint probability" of such multiple indexes as independent events analogically.* Pareto optimality, can only give a set of solution instead of the real optimized solution [1–5].

Besides, owing to the employment of personal factors, the relevant algorithms cannot be considered as full-quantitative methods in some sense. Meanwhile, the format of normalization and the selection of scaled factor (denominator) in their normalization process are problematic and troublesome issues, different normalization algorithms usually lead to different results [7]. The assessment of weighting factors is also not reasonable [8–11].

This situation indicates that an appropriate expression of "*simultaneous optimization of multiple objectives*" in a system is required urgently, so that a quantitatively overall optimization can be conducted.

The simultaneity of multiple objectives in optimization in a system reflects its integrity, irreplacement and inseparatable of each objective of the system. So, every goal is an constitutively organic component of the system. Dealing with the "simultaneous optimization" of multi-objective optimization problem is to optimize it as a whole, that is, to make the system in its best state as a whole, and all parts of it are "united" under the banner of "overall optimization" and "working" in coordination style [12–15]. According to the viewpoint of systems theory, "the whole system is not equal to the simple sum or superposition of its components", while "the whole is greater than the sum of its parts".

Therefore, it is necessary to find the "intersection" among the objectives, which is able to arrange these objectives "coordinated" and "harmonized" with each other, so that the optimization of the overall function of the system is achieved.

6.2 Probabilistic Multi-objective Optimization

The scientific concept of "intersection" comes from "set theory", which refers to a special set C that consists of all elements belonging to both set A and set B. Besides, C is recorded as AB or $A \cap B$.

Moreover, in term of "probability theory", the probability $P(C) \equiv P(AB)$ of two independent events A and B is equal to $P(A) \times P(B)$, i.e., $P(AB) = P(A) \times P(B)$, which is called the "joint probability" of the two independent events A and B, and represents the "simultaneous occurrence" of the two independent events A and B [16]; $P(A)$ and $P(B)$ are the corresponding probabilities of these two independent events A and B, respectively.

Therefore, the problem of "simultaneous optimization of multiple objectives" becomes a probabilistic problem of "simultaneous occurrence of multiple events" as each objective (attribute) in multi-objective optimization being regarded as an event. Furthermore, if each event is independent, the joint probability of the "overall optimization of the system" problem of "multiple events appearring at the same time" can be written as,

$$P_t = P_1 \cdot P_2 \cdot \ldots \cdot P_j \cdot \ldots = \prod_{j=1}^{m} P_j. \tag{6.1}$$

In Eq. (6.1), P_j expresses the occuring probability of j-th objective, and m indicates the total number of objectives. In this way, it thus transforms the optimization problem of multiple objectives into an equivalent joint probability problem.

6.3 Idea of Preferable Probability and Its Quantization

Take the material selection as an example [1].

In general, in a MOO problem some attribute utility indexes are with the characteristics of "*the higher the better*" [1, 2], i.e., the attribute with higher value is inevitably more welcomed and get more preference, this type of attribute is called *beneficial index*. In this case, a term "*preferable probability*" could be introduced to characterize the "*preference degree*" of the attribute in the optimization process reasonably [1, 2]. For the simplicity, it assumes that the *partial preferable probability* of such kind of attribute responses is proportional to the specific utility value of its attribute index positively, i.e.,

$$P_{ij} = \alpha_j X_{ij}, \quad i = 1, 2, \ldots, n, \quad j = 1, 2, \ldots, m. \tag{6.2}$$

In Eq. (6.2), n is the total number of alternatives in the relevant candidate system; m represents the total number of attribute (objective) indicators of each alternative; X_{ij} is the utility value of the j-th attribute (objective) indicator of the i-th alternative; P_{ij} represents the partial preferable probability of the beneficial attribute index X_{ij}; α_j represents the normalization factor of the j-th beneficial attribute (objective) indicator.

In the light of general principle of probability theory [16], for the j-th material objective (attribute) index, it derives following expression for the normalization factor [1, 2],

$$\alpha_j = 1/(n\overline{X}_j). \tag{6.3}$$

In Eq. (6.3), $\overline{X}_j$ is the average utility value of the j-th attribute index in the attribute group involved.

On the other hand, in a MOO problem some attribute utility indexes are with the characteristics of "*the lower the better*" [1, 2], i.e., the attribute with lower value is more welcomed and get more preference, this type of attribute responses is called *unbeneficial index*. In this case, the partial preferable probability of the unbeneficial attribute is to be linearly related to its attribute utility index value negatively in the optimization process equivalently,

$$P_{ij} = \beta_j(X_{j\max} + X_{j\min} - X_{ij}), \quad i = 1, 2, \ldots, n, \ j = 1, 2, \ldots, m. \tag{6.4}$$

In Eq. (6.4), both $X_{j\min}$ and $X_{j\max}$ represent the minimum and maximum utility values of the attribute indicators in the j-th attribute (objective) group, respectively; β_j is the normalization factor of the j-th unbeneficial attribute index, which can be expressed as [1, 2],

$$\beta_j = 1/[n(X_{j\max} + X_{j\min}) - n\overline{X}_j]. \tag{6.5}$$

6.4 Outline of Probabilistic Multi-objective Optimization

Furthermore, according to the fundamentals of probability theory [16], the *total preferable probability* of the i-th alternative can be written as,

$$P_i = P_{i1} \cdot P_{i2} \cdot \ldots \cdot P_{ij} \cdot \ldots = \prod_{j=1}^{m} P_{ij}. \tag{6.6}$$

Obviously, in this assessment, the *total preferable probability of each alternative is the unique and decisive index in the optimization process.* This is the essence of probabilistic multi-objective optimization (PMOO). Besides, through above procedures and using Eqs. (6.2)–(6.6), the *multi-objective optimization problem is now transformed into a single-objective optimization one in term of total preferable probability* undoubtedly. The *optimum alternative corresponds to the candidate that is with the highest total preferable probability, which is the optimal result of this overall optimization.*

Moreover, if there is a weighting factor w_j for j-th attribute response, the total preferable probability of the i-th alternative can be expressed in following form [1, 2],

$$P_i = P_{i1}^{w_1} \cdot P_{i2}^{w_2} \cdot \ldots \cdot P_{ij}^{w_j} \cdot \ldots = \prod_{j=1}^{m} P_{ij}^{w_j}. \tag{6.7}$$

Many applications of above approach are performed [1, 2], which gave acceptable results and conformed to known, and indicated the reasonability of this approach. The analysis of independence of two objective responses was discussed in literature detailedly [1].

6.5 Simple Application Examples of PMOO

1. **Option for experimental instrument**

A unit plans to select an instrument for experiment, and there are *four alternative* candidates (I_1, I_2, I_3, and I_4) to be chosen. Each instrument is characterized by *five attributes*, i.e., price, reliability, appearance, life span and weight. The quantitative value corresponding to each attribute can be characterized numerically. Table 6.1 represents the quantified utility values of attribute of each alternative candidate, which are represented by A_1, A_2, A_3, A_4 and A_5, respectively.

Obviously, among these five attributes, the reliability, appearance and life span belong to the beneficial type of indicators, while the price and weight belong to the unbeneficial type of indicators. Table 6.2 gives the assessment consequences.

Table 6.2 indicates that the 2nd alternative candidate instrument (I_2) exhibits the highest total preferable probability comparatively, therefore the optimum selection is 2nd alternative candidate in the overall optimization. Besides, the order of the four alternative candidate products is: $I_2 > I_1 > I_3 > I_4$.

2. **A design of beam with lightest weight and minimum cost**

The design problem of a beam with lightest weight and minimum cost is as following.

As is shown in Fig. 6.1 [18], it needs to cut a rectangular beam from a circular tree trunk. The fundamental requirements are that: (1). the height x_1 of the beam is not allowed to exceed a specific value h; (2). its inertia moment is not less than a specific value w, but the height x_1 of the beam is in range of its one width to four

Table 6.1 Decision matrix for instrument selection

Alternative candidate	Attribute				
	Relative price, A_1	Reliability, A_2	Appearance, A_3	Life span, A_4	Relative weight, A_5
I_1	8	7	9	4	6
I_2	7	6	8	3	3
I_3	6	5	7	3	5
I_4	10	4	6	5	7

Table 6.2 Assessed results of alternative candidate instrument

Alternative candidate	Partial preferable probability					Total	
	P_{A_1}	P_{A_2}	P_{A_3}	P_{A_4}	P_{A_5}	$P_i \times 10^3$	Rank
I_1	0.2424	0.3182	0.3000	0.2500	0.2105	1.2179	2
I_2	0.2727	0.2727	0.2667	0.2500	0.3684	1.8269	1
I_3	0.3030	0.2273	0.2333	0.1875	0.2632	0.7929	3
I_4	0.1818	0.1818	0.2000	0.3125	0.1579	0.3262	4

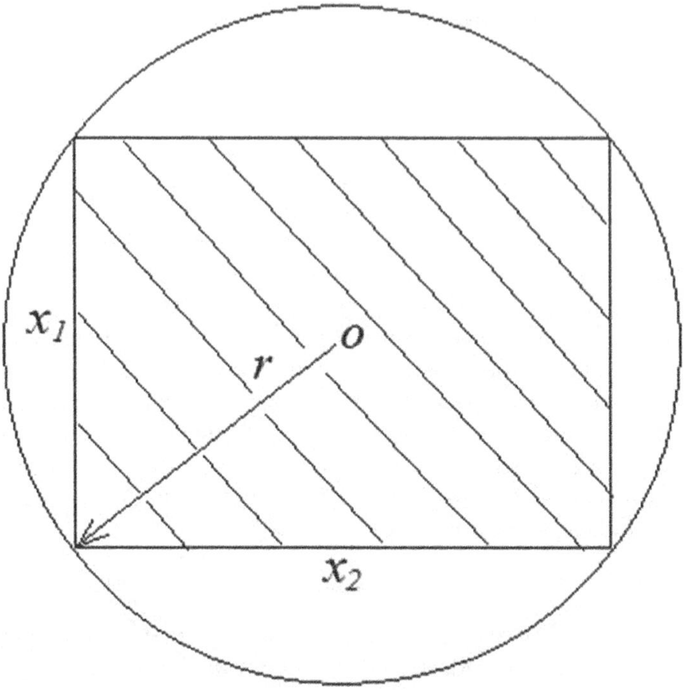

Fig. 6.1 Cutting a rectangular beam from a circular tree trunk

times width, i.e., $x_2 \le x_1 \le 4x_2$. Then, how to design it, so that the beam is with lightest weight and minimum cost.

Solution
As is shown in Fig. 6.1, x_1 and x_2 represent the height and width, respectively; r is the radius of circular tree trunk. Since the weight of the beam is proportional to its cross section area, i.e., $f_1 = x_1x_2$, so the first objective in the optimization, weight, of the beam can be written as,

$$\min f_1(x_1, x_2) = x_1 \cdot x_2. \tag{6.8}$$

While the second objective, cost of the tree trunk, is proportional to its cross section area of the tree trunk, $S = \pi r^2 = \pi\left[\left(\frac{x_1}{2}\right)^2 + \left(\frac{x_2}{2}\right)^2\right]$, so the cost objective of the trunk can be written as,

$$\min f_2(x_1, x_2) = x_1^2 + x_2^2. \tag{6.9}$$

Besides, the fundamental requirements (constraint conditions) of the design of beam with lightest weight and minimum cost can be written in following statements,

$$\text{s.t.}\begin{cases} g_1(x_1,x_2) = h - x_1 \geq 0; \\ g_2(x_1,x_2) = x_1^2 x_2 - w \geq 0; \\ g_3(x_1,x_2) : x_2 \leq x_1 \leq 4x_2; \\ g_4(x_1,x_2) : x_2 \geq 0; x_1 \geq 0. \end{cases} \tag{6.10}$$

Especially, it assumes $w = 1\ \text{m}^3$ and $h = 2.5$ m with the weighting factors of f_1 and f_2 of $\omega_1 = 0.3$ and $\omega_2 = 0.7$ [18], respectively. Thus it yields following specific fundamental requirements digitally,

$$\text{s.t.}\begin{cases} g_1(x_1,x_2) = 2.5 - x_1 \geq 0; \\ g_2(x_1,x_2) = x_1^2 x_2 - 1 \geq 0; \\ g_3(x_1,x_2) : 0.25x_1 \leq x_2 \leq x_1; \\ g_4(x_1,x_2) : x_1 \geq 0; x_2 \geq 0; \\ \omega_1 = 0.3; \omega_2 = 0.7. \end{cases} \tag{6.10'}$$

Figure 6.2 shows the requirements of g_1, g_2, g_3 and g_4 in diagram, respectively, which indicates that the effective area is the irregular quadrilateral zone *CEFG*.

By using quasi-uniform design $U_{64}(64^2)$ with 64 sampling points in rectangular area *ABCD* to conduct discretization visually, it obtains 19 sampling points falling in the irregular quadrilateral zone *CEFG*, which meets the basic requirement of point number (17) for sampling of uniform test design in two dimensional effective region [1], thus we could conduct the assessment with these 19 effective sampling points. A simple description on uniformly distributed effective sampling points will be given in next section additionally.

Table 6.3 represents the distributions of 64 sampling points in the rectangular area *ABCD* and the 19 effective sampling points falling in irregular quadrilateral zone *CEFG*, which are marked with label "*". Besides, the values of objective functions of f_1 and f_2, are presented in Table 6.3 as well.

Fig. 6.2 Requirements of g_1 through g_4 in diagram

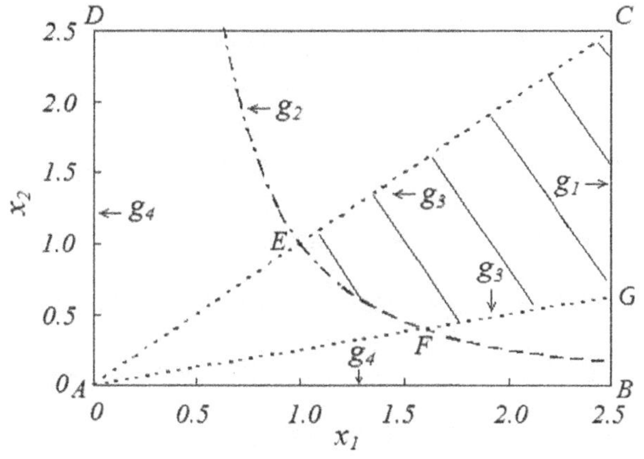

Table 6.3 Distributions of 64 sampling points and the 19 effective sampling points with quasi-uniform design $U_{64}(64^2)$

No	x_1/m	x_2/m	*CDEF*	f_1/m^2	f_2/m^2
1	0.019531	0.488281			
2	0.058594	0.996094			
3	0.097656	1.503906			
4	0.136719	2.011719			
5	0.175781	0.019531			
6	0.214844	0.527344			
7	0.253906	1.035156			
8	0.292969	1.542969			
9	0.332031	2.050781			
10	0.371094	0.058594			
11	0.410156	0.566406			
12	0.449219	1.074219			
13	0.488281	1.582031			
14	0.527344	2.089844			
15	0.566406	0.097656			
16	0.605469	0.605469			
17	0.644531	1.113281			
18	0.683594	1.621094			
19	0.722656	2.128906			
20	0.761719	0.136719			
21	0.800781	0.644531			
22	0.839844	1.152344			
23	0.878906	1.660156			
24	0.917969	2.167969			
25	0.957031	0.175781			
26	0.996094	0.683594			
27	1.035156	1.191406			
28	1.074219	1.699219			
29	1.113281	2.207031			
30	1.152344	0.214844			
31	1.191406	0.722656	*	0.8610	1.9417
32	1.230469	1.230469	*	1.5141	3.0281
33	1.269531	1.738281			
34	1.308594	2.246094			
35	1.347656	0.253906			
36	1.386719	0.761719	*	1.0563	2.5032

(continued)

Table 6.3 (continued)

No	x_1/m	x_2/m	*CDEF*	f_1/m^2	f_2/m^2
37	1.425781	1.269531	*	1.8101	3.6446
38	1.464844	1.777344			
39	1.503906	2.285156			
40	1.542969	0.292969			
41	1.582031	0.800781	*	1.2669	3.1441
42	1.621094	1.308594	*	2.1214	4.3404
43	1.660156	1.816406			
44	1.699219	2.324219			
45	1.738281	0.332031			
46	1.777344	0.839844	*	1.4927	3.8643
47	1.816406	1.347656	*	2.4479	5.1155
48	1.855469	1.855469	*	3.4428	6.8855
49	1.894531	2.363281			
50	1.933594	0.371094			
51	1.972656	0.878906	*	1.7338	4.6638
52	2.011719	1.386719	*	2.7897	5.97
53	2.050781	1.894531	*	3.8853	7.795
54	2.089844	2.402344			
55	2.128906	0.410156			
56	2.167969	0.917969	*	1.9901	5.5428
57	2.207031	1.425781	*	3.1467	6.9038
58	2.246094	1.933594	*	4.343	8.7837
59	2.285156	2.441406			
60	2.324219	0.449219			
61	2.363281	0.957031	*	2.2617	6.501
62	2.402344	1.464844	*	3.5191	7.917
63	2.441406	1.972656	*	4.8161	9.8518
64	2.480469	2.480469	*	6.1527	12.305

Table 6.4 shows the assessment results of the 19 effective sampling points. In the assessment, the objective functions of f_1 and f_2, all belong to unbeneficial indexes to contribute to their partial preferable probabilities, P_{f_1} and P_{f_2}, respectively; while the total preferable probability P_t is evaluated by using $P_t = P_{f_1}^{\omega_1} \cdot P_{f_2}^{\omega_2}$ with weighting factors of ω_1 and ω_2 [1]. The assessed results in Table 6.4 indicate that the sampling point No. 31 gains the maximum total preferable probability, so it is the optimum scheme of this design. As to the sample point *No.* 31, the input variables are $x_1{'} = 1.1914$ m and $x_2{'} = 0.7227$ m, respectively, the optimum objectives get their corresponding values of $f_1' = 0.8610$ m^2, $f_2' = 1.9417$ m^2, respectively.

Table 6.4 Assessment results of the 19 effective sampling points with quasi-uniform design $U_{64}(64^2)$

No	x_1/m	x_2/m	f_1/m^2	f_2/m^2	$P_{f1} \times 10$	$P_{f2} \times 10$	$P_t \times 10^2$	Rank
31	1.191406	0.722656	0.8610	1.9417	0.7448	0.7691	0.7617	1
32	1.230469	1.230469	1.5141	3.0281	0.6657	0.7012	0.6904	4
36	1.386719	0.761719	1.0563	2.5032	0.7212	0.7340	0.7301	2
37	1.425781	1.269531	1.8101	3.6446	0.6299	0.6627	0.6527	6
41	1.582031	0.800781	1.2669	3.1441	0.6957	0.6940	0.6945	3
42	1.621094	1.308594	2.1214	4.3404	0.5922	0.6192	0.6110	7
46	1.777344	0.839844	1.4927	3.8643	0.6683	0.6490	0.6547	5
47	1.816406	1.347656	2.4479	5.1155	0.5527	0.5707	0.5653	9
48	1.855469	1.855469	3.4428	6.8855	0.4323	0.4601	0.4516	14
51	1.972656	0.878906	1.7338	4.6638	0.6391	0.5990	0.6108	8
52	2.011719	1.386719	2.7897	5.9700	0.5113	0.5173	0.5155	11
53	2.050781	1.894531	3.8853	7.7950	0.3787	0.4033	0.3957	16
56	2.167969	0.917969	1.9901	5.5428	0.6081	0.5440	0.5625	10
57	2.207031	1.425781	3.1467	6.9038	0.4681	0.4590	0.4617	13
58	2.246094	1.933594	4.343	8.7837	0.3233	0.3415	0.3359	17
61	2.363281	0.957031	2.2617	6.5010	0.5752	0.4842	0.5098	12
62	2.402344	1.464844	3.5191	7.9170	0.4230	0.3956	0.4037	15
63	2.441406	1.972656	4.8161	9.8518	0.2660	0.2747	0.2721	18
64	2.480469	2.480469	6.1527	12.305	0.1042	0.1214	0.1159	19

3. **A simple description on uniformly distributed effective sampling points**

In literature [1], the discussions on uniformly distributed effective sampling points in spaces of one to three dimensions have been conducted in details in viewpoint of practical utilization. The approach of uniformly distributed effective sampling points is an efficient method to conduct space discretization for simplifing assessment in PMOO by means of good lattice point (GLP) and uniform experimental design (UED) visually. Here in this section, it gives simple demonstrations for practical and convenient applications additionally.

Early in 1959, Korobov put forward the idea of uniformly distributed "a point set". Thereafter Hua and Wang made a development of GLP method, which is with low–discrepancy in numerical integration in 1960s [1]. In the spirit of the GLP, the convergence speed of integration by using summation on GLP is much higher than that of the Monte Carlo (MC) sampling method. Later in 1980s, Fang and Wang formulated a UED method based on GLP or uniform sampling [1]. In the UED, the sampling points are uniformly distributed in the space, of which the positions are well deterministic instead of stochastically spreading. A reputation of "quasi–Monte Carlo method" (QMC) is given to such kind of algorithm thereafter [1].

A dramatic finding of good convergence speed of integration by using summation on uniformly distributed regular point sets, such as Halton sequences and Sobol sequences was found in 1990s in super space with very high dimensions even up to 360 dimensions, which implies that QMC methods exhibits very good convergence speed with respect to the convenient MC methods, and antithetic MC methods [1]. Papageorgiou et al. once reported that a simulation precision of order 10^{-2} was got with less than 500 points in 25 dimensional integral [1], and an empirical convergence rate approaching to n^{-1} was gained in their tests rather than the $n^{-1/2}$ of Monte Carlo, n is the sampling number of the uniformly distributed regular points.

Above results indicate the validity of QMC methods in simplifing integration by using summation on uniformly distributed regular point sets, which greatly promotes the development QMC methods.

The main features of uniform experimental design (UED) proposed by Fang and Wang proposed include:

(A) Uniformity

The sampling points are uniformly distributed in the variable space. UED spreads the sampling point regularly by using Fang's "Uniform Design Table".

(B) Overall Mean Model

UED expects that the sampling points supply the minimum deviation of the total mean value of the outcome response with respect to the actual total mean value.

(C) Robustness

UED can be used to a variety of condition with robustness despite of changes of model.

(D) Uniform Design Table

A series of "Uniform Design Table" and their Utility Tables were developed by Fang and Wang for the appropriate application of UED, which could be used to spread the sampling points uniformly with convenience.

As to the required number of uniformly distributed sampling points, there was detailed discussions in viewpoint of practical utilization [1].

In general, the value of a function (e.g., integrand), evolutes from point to point according to a certain regulation within a domain. An example is the sine function, which is represented as, $f(x) = A\cdot[1 + \sin(2\pi x/B)]$, in this expression A indicates the amplitude coefficient, B represents the period (wave length) of this sine function, and x reflects the coordinate value in one dimensional space.

It is clear that the peak value of this function is definitely $f = 2A$ at $x = x_0 = B/4$. While at a nearby position $x_1 = x_0 + \Delta x/2$, $\Delta x/2$ is the distance from the position x_0, the value of the function f becomes, $f_1 = A\cdot[1 + \sin(2\pi x_1/\mathrm{B})] = A\cdot\{1 + \sin[2\pi(x_0 + \Delta x/2)/B]\} = A\cdot[1 + \sin(\pi/2 + \Delta x\pi/B)]$.

Assume that a specific value $\Delta x\pi/B = 0.2856$ rad, the function f will take a particular value of $f_1 = 1.96\ A$ correspondingly, which leads to less deviation or error with respect to the mentioned peak value, 2 A [1].

Above analysis shows that if one attempts to provide an approximation value of the periodic function $f(x)$ at its peak value position by a nearby point with less error or deviation by sub–cutting the period, the partition number κ_1 of the sub–cutting in the period range (wave length) B for such periodic function within its single period is, $\kappa_1 = B/\Delta x = \pi/0.2856 \cong 11$. Correspondingly, the distance between the nearest neighbor sampling points is $\Delta x = B/11$. This indicates that the 11 sampling points in one single periodic range (wave length) could supply an efficient approximation to the peak value with less error or deviation to its peak value of the function in one dimension. Therefore, it could take 11 uniformly distributed sampling points for one dimensional case for practical application [1].

As to the case of two dimensional space, the problem becomes a plane issue where a rectangular coordinate system could be established, which consists of two orthogonal coordinate axes, says X and Y axes.

Preliminarily, the UED method states that the projections of any two sample points on one coordinate axis is not coincide. In the worst case, all the sampling points could be distributed along the diagonal line of the square. Thus, the distance between the nearest neighbor sampling points is enlarged by $\sqrt{1^2 + 1^2} = \sqrt{2}$ times as compared to that of the distance between the nearest neighbor sampling points of one dimensional case. Therefore, if one attempts to supply an appropriate approximation with a similar precision as that in one dimensional case for the function, the sub—cutting should be refined by an adjusting coefficient $1/\sqrt{2}$, here we take a factor 1/ 1.5, which leads to the number of sampling points κ_2 to the period (wave length) B range of this periodical function within one period to be $\kappa_2 = 1.5 \times \kappa_1 = 1.5 \times 11 = 16.5 \cong 17$. This result indicates that 17 sampling points for two dimensions in one single periodic range (wave length) could roughly provide an appropriate approximation for the peak value of the sine function with a similar precision like that in one dimension to its peak value for the sampling points along the diagonal line of the square.

Analogically, in three dimensional space, it is a cube in principle, a rectangular coordinate system could be established, which consists of three orthogonal coordinate axes, i.e., X, Y and Z axes in general. Again, considering the worst case that the distribution of all the sampling points are conducted along the diagonal line of the cube. The distance between the nearest neighbor sampling points is enlarged by $\sqrt{1^2 + 1^2 + 1^2} = \sqrt{3}$ times as compared to that of the distance between the nearest neighbor sampling points of one dimensional case. Therefore, if one tries to provide an appropriate approximation for the peak value of the function with a similar precision as that of the one dimensional case for the function once more, the sub-cutting is refined by an adjusting coefficient $1/\sqrt{3} \approx 1/1.732$, which results in the number of sampling points κ_3 to the period (wave length) B range of this periodical function within one period, $\kappa_3 = 1.732 \times \kappa_1 = 1.732 \times 11 = 19.052 \cong 19$. This indicates that the 19 sampling points of the one single periodic range (wave length) could provide an accurate approximation for the peak value of the sine function with a similar precision like that in one dimensional case to its peak value in three dimensions for the sampling points along the diagonal line of the cube assumedly.

Afterward, one could adjust the spatial distributions of the sampling points according to the second demand of the uniform design method [1].

In summary, it indicates that if one attempts to provide an appropriate estimation for a periodical function in its single peak domain, 11 sampling points in one dimensional case, 17 sampling points in two dimensional case, and 19 sampling points in three dimensional case are assumedly very necessary for the estimation of the peak value of the function, respectively; the sampling points must be uniformly distributed according to the rule of GLP and uniform design method deterministically. Many practical examples are provided to show the effectiveness of the finite number of uniformly distributed sampling point approach in literature [1].

6.6 Robust Design as a Special Case of Probabilistic Multi-objective Optimization

Robust design and assessment of product quality and production process are optimal problems in principle, which aim to reduce the effect of uncertain factors on the evaluated goals, so it is the specific case of probabilistic multi-objective optimization in principle [2, 17].

Early in 1950s, Box and Taguchi noticed the importance of robust design [19, 20]. The so called Taguchi method was proposed by Taguchi to seek the specific controllable parameters or conditions of production process, so that the quality responses are insensitive to the uncertain factors, such as environmental disturbance or surrounding factors, etc.

Taguchi put forward a term—"*signal to noise ratio* (*SNR*)" to assess robust design [20], but *its reasonability was questioned by statisticians due to the lack of unsatisfactory treatments* for both *mean value and variation of objective responses*, which actually *need to be considered separately* [1, 2, 21]. Subsequently, many modifications and improvements have been given to make improvements [22–32], but the intrinsic shortcomings remained [1, 2, 33].

In recent years, the PMOO considers the optimization problem of many objectives being the overall optimization of multiple objectives within a system [1, 2], which takes *every objective as one independent event of probability theory* analogically; thus the maximum of the *joint probability of the alternative in the system is the optimum status of the system* according to systems theory and probability theory [1, 2].

In the probabilistic robust design, both the *mean value and its dispersion value of each attribute are taken as dual sub-objectives* to conduct the optimization, thus a robust design problem becomes a *dual multi-objective optimization one.* The *mean value is the representative to exhibit the role of the attribute* in the assessment, while

Table 6.5 Designed scheme of parameter with $L_9(3^4)$

Parameter	Level		
	1	2	3
A: melt temperature (°C)	195	185	205
B: injection time (s)	15	12	18
C: cooling time (s)	40	35	45
D: pressure (MPa)	7.5	6.5	7.0

the *dispersion value of the attribute possesses the characteristic of unbeneficial type of attribute.*

1. **Application in case of "the smaller the better"**

As an example for "the smaller the better" case [1], the robust optimization design for minimum of injection warping deformation of ABS plastic product is presented here [34].

In an injection of plastic product process, warping deformation is employed as the response attribute to conduct the robust design. The robust design for such response of course has the feature "the lower the better", i.e., unbeneficial type of attribute. In the experiments of Yang et al. [34], an orthogonal array experiment with $L_9(3^4)$ was used. The controlling parameters include melt temperature (*A*), injection time (*B*), cooling time (*C*), and pressure (*D*), which are shown in Table 6.5.

Five times' tests for each experimental scheme of the nine designed schemes are conducted. Table 6.6 shows the test result of warping deformation from these designed experimental schemes.

In this robust design problem, less warping deformation is of course the optimal goal, so both the mean value of the warping deformation and its dispersion value belong to the unbeneficial type of performance indexes inevitably. The assessment

Table 6.6 Test result of warping deformation from the designed scheme

No	Input parameter				Warping deformation, mm					Representative	
	A	*B*	*C*	*D*	y_1	y_2	y_3	y_4	y_5	Mean, $\bar{y}$, mm	Deviation, σ, mm
1	1	1	1	1	0.68	0.72	0.68	0.70	0.72	0.7	0.0179
2	1	2	2	2	0.68	0.74	0.70	0.70	0.74	0.712	0.0240
3	1	3	3	3	0.72	0.72	0.76	0.68	0.74	0.724	0.0265
4	2	1	2	3	0.66	0.74	0.70	0.72	0.72	0.708	0.0271
5	2	2	3	1	0.62	0.66	0.64	0.64	0.62	0.636	0.0150
6	2	3	1	2	0.64	0.68	0.66	0.68	0.62	0.656	0.0233
7	3	1	3	2	0.64	0.64	0.60	0.60	0.62	0.620	0.0179
8	3	2	1	3	0.68	0.78	0.74	0.76	0.76	0.744	0.0344
9	3	3	2	1	0.60	0.60	0.60	0.62	0.60	0.604	0.0080

results are shown in Table 6.7. The labels $P_{\overline{y}}$ and P_{σ} in Table 6.7 present parts of partial preferable probability of the mean value of warping deformation and its deviation value, respectively; P_t reflects the entire partial (here in fact the total as well) preferable probability of each scheme, which uniquely determines the ranking of each scheme in Table 6.7.

Table 6.7 reveals that scheme No. 9 is the optimum configuration, which is with both mean value of warping deformation and its deviation value overall minimum as a whole comprehensively.

2. **Application in case of "the larger the better"**

As an example for "the larger the better" case [1], the robust optimization design for maximum of an elastometric connector is presented [35].

In an elastometric connector design process, it is to assemble an elastometric connector to a nylon tube, so the design objective is to maximize the pull-off force. The robust design for such response of course has the feature "the larger the better", i.e., beneficial type of attribute. In the experiments of Byrne et al. [35], an orthogonal array experiment with $L_9(3^4)$ was used. Four controlling parameters are involved, which include interference (A), connector wall thickness (B), insertion depth (C), and percent adhesive in connector pre-dip (D). Table 6.8 shows the parameter design of experiments.

Table 6.7 Assessed results of the preferable probability of all schemes and their ranking for warping deformation robust design

No	$P_{\overline{y}}$	P_{σ}	$P_t \times 10^2$	Rank
1	0.1075	0.1307	1.4055	3
2	0.1055	0.0982	1.0357	6
3	0.1035	0.0847	0.8763	7
4	0.1062	0.0815	0.8650	8
5	0.1181	0.1463	1.7283	2
6	0.1148	0.1018	1.1682	4
7	0.1208	0.1307	1.5790	5
8	0.1002	0.0427	0.4274	9
9	0.1234	0.1835	2.2645	1

Table 6.8 Factors and levels of elastometric connector and tube

Controllable factor	Level		
	1	2	3
Interference, A	Low	Medium	High
Wall thickness, B	Thin	Medium	Thick
Insertion depth, C	Shallow	Medium	Deep
Percent adhesive, D	Low	Medium	High

Eight times' tests for each experimental scheme of the nine designed schemes are conducted. Table 6.9 shows the test result of pull-off force from these designed experimental schemes.

The assessment results are shown in Table 6.10. The labels $P_{\bar{y}}$ and P_σ in Table 6.10 present parts of partial preferable probability of the mean value of pull-off force and its deviation value, respectively; P_t reflects the entire partial (here in fact the total as well) preferable probability of each scheme.

Table 6.10 reveals that scheme *No.* 4 is the optimum configuration, which is with both mean value of pull-off force maximum and its deviation value overall minimum as a whole comprehensively.

Table 6.9 Test result of pull-off force for designed experimental schemes with $L_9(3^4)$

No	Input parameter				Pull-off force, N								Representative	
	A	*B*	*C*	*D*	y_1	y_2	y_3	y_4	y_5	y_6	y_7	y_8	$\bar{y}$, N	σ, N
1	1	1	1	1	15.6	9.5	16.9	19.9	19.6	19.6	20.0	19.1	17.525	3.3793
2	1	2	2	2	15.0	16.2	19.4	19.2	19.7	19.8	24.2	21.9	19.425	2.7197
3	1	3	3	3	16.3	16.7	19.1	15.6	22.6	18.2	23.3	20.4	19.025	2.6971
4	2	1	2	3	18.3	17.4	18.9	18.6	21.0	18.9	23.2	24.7	20.125	2.4299
5	2	2	3	1	19.7	18.6	19.4	25.1	25.6	21.4	27.5	25.3	22.825	3.2061
6	2	3	1	2	16.2	16.3	20.0	19.8	14.7	19.6	22.5	24.7	19.225	3.1614
7	3	1	3	2	16.4	19.1	18.4	23.6	16.8	18.6	24.3	21.6	19.850	2.7920
8	3	2	1	3	14.2	15.6	15.1	16.8	17.8	19.6	23.2	24.2	18.313	3.4884
9	3	3	2	1	16.1	19.9	19.3	17.32	23.1	22.7	22.6	28.6	21.204	3.6903

Table 6.10 Assessed results of the preferable probability of all schemes and their ranking for elastometric connector robust design

No	$P_{\bar{y}}$	P_σ	$P_t \times 10^2$	Rank
1	0.0987	0.0996	0.9834	9
2	0.1094	0.1236	1.3522	4
3	0.1072	0.1244	1.3332	5
4	0.1134	0.1341	1.5204	1
5	0.1286	0.1059	1.3616	2
6	0.1083	0.1075	1.1645	6
7	0.1118	0.1209	1.3525	3
8	0.1032	0.0956	0.9866	8
9	0.1194	0.0883	1.0547	7

6.7 Fuzzification as an Auxiliary Measure for PMOO Robust Design in Case of Desired Value as Target

1. **Necessity**

Proper parameter design of production process is significant to guarantee quality improvement of products. As to robust design, the *mean value* and *its variance of an objective are taken as dual sub-objectives of the multi-objective optimization problem*, thus the robust design problem seems to be solved. Under conditions of the "smaller the better" and "the larger the better", it is no doubt to solve the problems as they are. Currently, for the case of desired "target the best", the departure of the actually tested values from the target value is considered [2, 33], it is a promising solution, especially the membership approach in fuzzification of mean value of tested result belonging to the desired target value of an objective response is taken as one objective response, which is a brand new method and value to be examined extensively. Under some conditions, fuzzification is more convenient to get an appropriate consequence and diminish erroneous flexibly [1, 2].

However, since the maximum value μ of membership is 1, a definite value, instead of infinite, so an alternative appropriate manner to deal with this problem can be put forward by introducing a "complement" of the membership value, i.e., $\eta = 1-\mu$, to conduct the evaluation logically. Additionally, in condition of robust assessment, the dispersion of test data must be taken into account in proper manner.

In this section, a rational robust parameter design of production in case of desired value as target is developed in term of "complement" of the membership value as an auxiliary measure; moreover, three examples are represented to show the procedure, which includes gas metal arc (GMA) welding process, parametric design of printing machine's ability, and robust parametric design for coating thickness of bare silicon wafers quality controlling.

2. **Regulation of fuzzied probabilistic robust design in case of desired value as target**

2.1 **Combination of PMOO with membership approach in fuzzification**

In case of desired value as target, the actual closeness of each test data y to the desired target value y_0 can be employed to measure its closeness to the desired target, thus the greater the distance of the test data y from the desired target value y_0 is, the smaller the closeness or degree of proximity to the desired target. This is something like the membership in fuzzy theory [2]. The assessment of membership of a test data y belonging to the desired target value y_0 rationally is,

$$\begin{cases} \mu = 1, y = y_0; \\ \mu = 1 - |y - y_0|/\delta, |y - y_0| \le \delta; \\ \mu = 0, |y - y_0| > \delta. \end{cases} \tag{6.11}$$

In Eq. (6.11), μ is the membership function of the test data y belonging to the desired target value y_0, δ is the pre-assigned value, beyond which the value of membership function becomes 0.

Since the limit value of membership of y belonging to y_0 is "1", which is a definite value instead of infinitely large. So, it seems improper to consider this optimization problem as "the larger the better" type, i.e., beneficial indicator, since in the latter type the value of the objective response has the possibility to take a value of infinitely large value instead of finite one.

Alternatively, the "complement" η of the membership value can be used as the appropriate index to deal with this problem by introducing it with Eq. (6.12),

$$\eta = 1 - \mu. \tag{6.12}$$

The limit value of η is 0, which is corresponding to μ taking its maximum value of 1. Thus, the optimization problem of μ taking its maximum value is transformed into an equivalent description of η approaching to its minimum value of 0.

Furthermore, as to robustness assessment, since the unavoidability of dispersion of a set of test data at the same experimental conditions due to the effects of external uncertain factors, the evaluation of dispersion of a set of test data must be taken into account surely [1, 2, 21, 33].

In the light of Lin and Tu's discussion [2, 24, 33], the dispersion of a set of test data in term of membership of fuzzy theory can be characterized as follows,

$$s_\mu = (\overline{\eta}^2 + \sigma_\mu^2)^{0.5}. \tag{6.13}$$

In Eq. (6.13), σ_μ is the standard error of membership value μ of the set of test data at the same experimental conditions; $\overline{\eta}$ is the mean value of "complement"η of the membership value in the corresponding test data set; s_μ is in fact the dispersion of a set of test data in term of membership with respect to the desired target value to join the assessment of the second part of partial preferable probability; while $\overline{\eta}$ is an unbeneficial index to join the assessment of the first part of partial preferable probability.

2.2 Assessment of preferable probability

Furthermore, the parts of partial preferable probability of both s_μ and $\overline{\eta}$ of an objective, P_{s_μ} and $P_{\overline{\eta}}$, can be conducted as unbeneficial type of indicators to perform the assessment [1, 2, 21, 33]. As a result, the partial preferable probability P_{ij} of the j-th objective of i-th candidate scheme (alternative) is the product of both P_{s_μ} and $P_{\overline{\eta}}$. Subsequently, the total preferable probability P_i of i-th candidate scheme is [1, 2, 33],

$$P_i = \prod_{j=1}^{m} P_{ij}, \quad \mathrm{i} = 1, 2, \ldots, \mathrm{n}; \quad \mathrm{j} = 1, 2, \ldots, \mathrm{m}. \tag{6.14}$$

Finally, according to probability-based multi-objective optimization (PMOO) [1, 2, 33], the optimal option in the system is the specific scheme that is with the highest total preferable probability.

3. **Applications**

3.1 **Application in GMA welding process**

As an application of the robust parameter optimization in case of desired value as target, the parameters optimization for gas metal arc (GMA) welding process is employed to illuminate the procedure in detail first.

Kim and Rhee once studied the GMA welding process with dual response approach [36]. Here in this section, this problem is restudied by using the newly developed probabilistic robust design of product to demonstrate the utilization of fuzzification treatment in case of desired value as target.

The base metal of the welding material in Kim and Rhee's study was mild steel [36], which was to be welded as I-groove typed joint with the thickness of 5.8 mm. The AWS ER 70S-6 brand electrode wire was used, of which is the diameter is 1.2 mm. CO_2 was used as shielding gas which flows at a rate of 20 L/min. A constant voltage mode—welding power source was used. The variation range of the root opening was in 0.4–1.2 mm [36]. The desired target value for penetration y_0 is 3.5 mm.

In Kim and Rhee's study, 9 groups of tests were conducted according to the experimental design. The mean value and standard deviation of the output response were then caused from the test data of these 5 specimens for each experimental design. The input controllable variables include the rate (x_1) of wire-feed and welding rate (x_2). The output response is the penetration, which consisted of the mean value $\overline{y}$ and standard error σ of the penetration. The level of each input controllable variable is citied and shown in Table 6.11. Their experimental results are cited in Table 6.12 with 2^3 factorial design.

The evaluation consequences for the membership values μ and the corresponding errors of GMA welding process are shown in Table 6.13, the pre-assigned value of δ is chosen as $\delta = 1$ mm.

Table 6.14 shows that the test scheme No. 2 gains the maximum total preferable probability P_t, so the robust status from these test data is the test No. 2 for the system, which is with the mean value of penetration of 3.42 mm and standard error of $\sigma = \left[\sum_{i=1}^{5} (f_i - \overline{f})^2/5\right]^{0.5} = 0.3487$ mm at wire-feed rate of $x_1{}^* = 75$ mm/s and welding rate of $x_2{}^* = 6$ mm/s.

3.2 **Application in parametric design of printing machine's ability**

Table 6.11 Parameter design of GMA welding process experiment

Input variable	Level 1	Level 2	Level 3
Rate of wire feed (mm/s), x_1	60	75	90
Welding rate (mm/s), x_2	6	8	10

Table 6.12 Experimental results of GMA welding process with 2^3 factorial design

No	Input variable		Experimental result of penetration f, mm					Mean value, mm	Standard error, mm
	x_1, (mm/s)	x_2, (mm/s)	f_1	f_2	f_3	f_4	f_5	$\overline{f}$	σ
1	60	6	2.7	2.9	3.1	3.1	3.4	3.04	0.2332
2	**75**	**6**	**3.0**	**3.2**	**3.3**	**3.6**	**4.0**	**3.42**	**0.3487**
3	90	6	3.6	3.8	4.2	4.3	4.5	4.08	0.3311
4	60	8	2.4	2.6	2.6	2.7	2.8	2.62	0.1327
5	75	8	2.5	2.7	3.0	3.5	3.8	3.10	0.4858
6	90	8	2.9	3.3	3.6	3.7	4.1	3.52	0.4020
7	60	10	1.9	1.9	2.2	2.2	2.5	2.14	0.2245
8	75	10	1.8	2.4	2.7	2.9	3.2	2.60	0.4775
9	90	10	2.6	2.7	3.2	3.5	3.9	3.18	0.4874

Table 6.13 Membership values μ and the corresponding errors of GMA with $\delta = 1$ mm

No	Membership function μ					Mean value $\overline{\mu}$	$\overline{\eta}$	σ_{μ}	s_{μ}
1	0.2	0.4	0.6	0.6	0.9	0.54	0.46	0.2332	0.5158
2	0.5	0.7	0.8	0.9	0.5	0.68	0.32	0.1600	0.3578
3	0.9	0.7	0.3	0.2	0	0.42	0.58	0.3310	0.6678
4	0	0.1	0.1	0.2	0.3	0.14	0.86	0.1020	0.8660
5	0	0.2	0.5	1	0.7	0.48	0.52	0.3544	0.6293
6	0.4	0.8	0.9	0.8	0.4	0.66	0.34	0.2154	0.4025
7	0	0	0	0	0	0	1	0	1
8	0	0	0.2	0.4	0.7	0.26	0.74	0.2653	0.786
9	0.1	0.2	0.7	1	0.6	0.52	0.48	0.3311	0.5831

Table 6.14 Evaluation results of the preferable probability

No	$P_{\overline{\eta}}$	$P_{s_{\mu}}$	$P_t \times 10^2$	Rank
1	0.1307	0.1277	1.6685	3
2	**0.1520**	**0.1478**	**2.2466**	**1**
3	0.1125	0.1082	1.2173	6
4	0.0699	0.0829	0.5798	8
5	0.1216	0.1132	1.3759	5
6	0.1489	0.1421	2.1166	2
7	0.0486	0.0658	0.3202	9
8	0.0881	0.0931	0.8210	7
9	0.1277	0.1191	1.5199	4

Vining and Myers once raised the problem of parametric design of printing machine's ability [37, 38]. Three factors were involved, i.e., speed x_1, pressure x_2 and distance x_3, which affect the printing machine's ability to use ink to package labels. A 3^3 complete factorial was used in the experiments which were with three replicates at each design point. The goal was to seek the optimal location at which the printing number is around 500 with minimum variance. Table 6.15 cited the experimental results.

From Table 6.15, it can be seen that the sample standard deviation with zero value at two locations, i.e., alternatives *No.* 10 and *No.* 14, but their printing numbers

Table 6.15 Cited experimental result of printing ink data

No	Coded input variable			Output response			Mean value	Standard error
	x_1	x_2	x_3	y_1	y_2	y_3	$\bar{y}$	σ
1	− 1	− 1	− 1	34	10	28	24.00	12.49
2	0	− 1	− 1	115	116	130	120.33	8.39
3	1	− 1	− 1	192	186	263	213.67	42.83
4	− 1	0	− 1	82	88	88	86.00	3.46
5	0	0	− 1	44	178	188	136.67	80.41
6	1	0	− 1	322	350	350	340.67	16.17
7	− 1	1	− 1	141	110	86	112.33	27.57
8	0	1	− 1	259	251	259	256.33	4.62
9	1	1	− 1	290	280	245	271.67	23.63
10	− 1	− 1	0	81	81	81	81.00	0.00
11	0	− 1	0	90	122	93	101.67	17.67
12	1	− 1	0	319	376	376	357.0	32.91
13	− 1	0	0	180	180	154	171.33	15.01
14	0	0	0	372	372	372	372.0	0.00
15	1	0	0	541	568	396	501.67	92.50
16	− 1	1	0	288	192	312	264.00	63.50
17	0	1	0	432	336	513	427.00	88.61
18	1	1	0	713	725	754	730.67	21.08
19	− 1	− 1	1	364	99	199	220.67	133.82
20	0	− 1	1	232	221	266	239.67	23.46
21	1	− 1	1	408	415	443	422.00	18.52
22	− 1	0	1	182	233	182	199.00	29.44
23	0	0	1	507	515	434	485.33	44.64
24	1	0	1	846	535	640	673.67	158.21
25	− 1	1	1	236	126	168	176.67	55.51
26	0	1	1	660	440	403	501.00	138.94
27	1	1	1	878	991	1161	1010.00	142.45

are far from the desired target value for number obviously. Table 6.16 provides the fuzzification evaluation results with $y_0 = 500$ and $\delta = 200$. Table 6.17 shows the evaluation results by means of PMOO, the alternative No. 23 of experimental scheme gains the highest overall preferable probability, which can be selected as the optimal design for the system. At scheme No. 23, i.e., coded input variables speed $x_1{}^* = 0$, pressure $x_2{}^* = 0$ and distance $x_3{}^* = 1$, the mean printing number is 485.33 with standard error of 44.64.

This approach avoids the effect of the puzzled value like the sample standard deviation of zero, such as in this example at two locations, alternatives No. 10 and No. 14.

Table 6.16 Fuzzification evaluation results with $y_0 = 500$ and $\delta = 200$

No	Membership function μ			Mean value $\overline{\mu}$	σ_μ	$\overline{\eta}$	s_μ
1	0	0	0	0	0	1	1
2	0	0	0	0	0	1	1
3	0	0	0	0	0	1	1
4	0	0	0	0	0	1	1
5	0	0	0	0	0	1	1
6	0.11	0.25	0.25	0.2033	0.0660	0.7967	0.7994
7	0	0	0	0	0	1	1
8	0	0	0	0	0	1	1
9	0	0	0	0	0	1	1
10	0	0	0	0	0	1	1
11	0	0	0	0	0	1	1
12	0.095	0.38	0.38	0.2850	0.1344	0.7150	0.7275
13	0	0	0	0	0	1	1
14	0.36	0.36	0.36	0.3600	0	0.6400	0.6400
15	0.795	0.66	0.48	0.6450	0.1290	0.3550	0.3777
16	0	0	0.06	0.0200	0.0283	0.9800	0.9804
17	0.66	0.18	0.935	0.59107	0.3120	0.4083	0.5139
18	0	0	0	0	0	1	1
19	0.32	0	0	0.1067	0.1508	0.8933	0.9060
20	0	0	0	0	0	1	1
21	0.54	0.575	0.715	0.6100	0.0756	0.39	0.3973
22	0	0	0	0	0	1	1
23	**0.965**	**0.925**	**0.67**	**0.8533**	**0.1307**	**0.1467**	**0.1964**
24	0	0.825	0.3	0.37	0.3410	0.6250	0.7120
25	0	0	0	0	0	1	1
26	0.2	0.7	0.515	0.4717	0.2064	0.5283	0.5672
27	0	0	0	0	0	1	1

Table 6.17 Evaluation results by means of PMOO

No	$P_{\overline{\eta}}$	$P_{s_{\mu}}$	$P_t \times 10^4$	Rank
1	0.0252	0.0263	6.6100	
2	0.0252	0.0263	6.6100	
3	0.0252	0.0263	6.6100	
4	0.0252	0.0263	6.6100	
5	0.0252	0.0263	6.6100	
6	0.0396	0.0402	15.9170	9
7	0.0252	0.0263	6.6099	
8	0.0252	0.0263	6.6099	
9	0.0252	0.0263	6.6099	
10	0.0252	0.0263	6.6099	
11	0.0252	0.0263	6.6099	
12	0.0454	0.0452	20.5129	8
13	0.0252	0.0263	6.6099	
14	0.0507	0.0513	26.0012	6
15	0.0709	0.0695	49.2941	2
16	0.0262	0.0276	7.3444	
17	0.0671	0.0601	40.3112	4
18	0.0252	0.0263	6.6100	
19	0.0327	0.0328	10.7356	10
20	0.0252	0.0263	6.6100	
21	0.0684	0.0682	46.6394	3
22	0.0252	0.0263	6.6100	
23	**0.0857**	**0.0821**	**70.3624**	**1**
24	0.0517	0.0463	23.9576	7
25	0.0252	0.0263	6.6100	
26	0.0586	0.0564	33.0277	5
27	0.0252	0.0263	6.6100	

3.3 Robust parametric design for coating thickness of bare silicon wafers quality controlling

Shin et al. studied the robust parametric design for coating thickness of bare silicon wafers quality controlling [39]. Three main parameters that might influence the coating thickness of bare silicon wafers were selected, i.e., mould temperature x_1, injection flow rate x_2, and cooling rate x_3. The spherical central composite designed parameter of the three control factors is cited and shown in Table 6.18. The test results are cited and represented in Table 6.19. The desired target value of the coating thickness is 71.14 μm. If we set $\delta = 10$ μm, the evaluation results for μ and η are shown in Table 6.20. Furthermore, the probabilistic evaluation consequences and ranking are shown in Table 6.21.

The results in Table 6.21 indicates that the top 3 alternatives are schemes No. 15, No. 17 and No. 19, respectively, which are with close total preferable probability. If scheme No. 15 is selected as the optimum status for the system, the corresponding

Table 6.18 Parameter design of three control factors with spherical central composite design

Levels of coded variables	Mould temperature (F), x_1	Injection flow rate (lbs/s), x_2	Cooling rate (F/s), x_3
− α (−1.682)	99.6	40	4.8
− 1	140.5	46	15
0	200.5	55	30
+ 1	260.5	64	45
+ α (+ 1.682)	301.4	70	55.2

Table 6.19 Test results with spherical central composite design

No	Input variable			Coating thickness of bare silicon wafers (μm)				Mean value (μm)
	x_1	x_2	x_3	y_1	y_2	y_3	y_4	$\bar{y}$
1	− 1	− 1	− 1	76.3	80.5	77.7	81.1	78.90
2	1	− 1	− 1	79.1	81.2	78.8	79.6	79.68
3	− 1	1	− 1	82.5	81.5	79.5	80.9	81.10
4	1	1	− 1	72.3	74.3	75.7	72.7	73.75
5	− 1	− 1	1	70.6	72.7	69.9	71.5	71.18
6	1	− 1	1	74.1	77.9	76.2	77.1	76.33
7	− 1	1	1	78.5	80.0	76.2	75.3	77.50
8	1	1	1	84.9	83.1	83.9	83.5	83.85
9	− 1.682	0	0	74.1	71.8	72.5	71.9	72.58
10	1.682	0	0	76.4	78.7	79.2	79.3	78.40
11	0	− 1.682	0	79.2	80.7	81.0	82.3	80.80
12	0	1.682	0	77.9	76.4	76.9	77.4	77.15
13	0	0	− 1.682	82.4	82.7	82.6	83.1	82.70
14	0	0	1.682	79.7	82.4	81.0	81.2	81.08
15	0	0	0	70.4	70.6	70.8	71.1	70.73
16	0	0	0	70.9	69.7	69.0	69.9	69.88
17	0	0	0	70.7	71.9	71.7	71.2	71.38
18	0	0	0	70.2	71.0	71.5	70.4	70.78
19	0	0	0	71.5	71.1	71.2	70.0	70.95
20	0	0	0	71.0	70.4	70.9	69.9	70.55

Table 6.20 Evaluation results for μ and η

No	Membership of y belonging to 71.14 μm					Complement		
	μ_1	μ_2	μ_3	μ_4	$\overline{\mu}$	$\overline{\eta}$	σ_μ	s_μ
1	0.484	0.064	0.344	0.004	0.224	0.776	0.1975	0.8007
2	0.204	0	0.234	0.154	0.148	0.852	0.0901	0.8568
3	0	0	0.164	0.024	0.047	0.953	0.0683	0.9554
4	0.884	0.684	0.544	0.844	0.739	0.261	0.1352	0.2939
5	0.946	0.844	0.876	0.964	0.9075	0.0925	0.0492	0.1048
6	0.704	0.324	0.494	0.404	0.4815	0.5185	0.1418	0.5376
7	0.264	0.114	0.494	0.584	0.364	0.636	0.1856	0.6625
8	0	0	0	0	0	1	0	1
9	0.704	0.934	0.864	0.924	0.8565	0.1435	0.0920	0.1705
10	0.474	0.244	0.194	0.184	0.274	0.726	0.1177	0.7355
11	0.194	0.044	0.014	0	0.063	0.937	0.0773	0.9402
12	0.324	0.474	0.424	0.374	0.399	0.601	0.0559	0.6036
13	0	0	0	0	0	1	0	1
14	0.144	0	0.014	0	0.0395	0.9605	0.0606	0.9624
15	0.926	0.946	0.966	0.996	0.9585	0.0415	0.0259	0.0489
16	0.976	0.856	0.786	0.876	0.8735	0.1265	0.0680	0.1436
17	0.956	0.924	0.944	0.994	0.9545	0.0455	0.0255	0.0522
18	0.906	0.986	0.964	0.926	0.9455	0.0545	0.0313	0.0629
19	0.964	0.996	0.994	0.886	0.96	0.04	0.0446	0.0599
20	0.986	0.926	0.976	0.876	0.941	0.059	0.0439	0.0735

input parameters are, mould temperature at $x_1* = 200.5$ F, injection flow rate $x_2* = 55$ lbs/s, and cooling rate $x_3* = 30$ F/s, respectively.

6.8 Reliability Design Involving Optimization with Multiple Objectives by Means of Probabilistic Multi-objective Optimization

For the reliability optimization design of mechanical parts, there are many achievements on single-objective model. However, the reliability research involving multi-objective optimization problems has not obtained satisfactory research results yet. In the past, the main difficulty of multi-objective reliability optimization design of mechanical parts was due to the lack of systemic viewpoint and appropriate method in solving the overall optimization of multi-objective optimization problem in a system.

Table 6.21 Probabilistic evaluation consequences and ranking

No	P_η	$P_{s\mu}$	$P_t \times 10^3$	Rank
1	0.0241	0.0227	0.5469	14
2	0.0171	0.0176	0.3016	15
3	0.0079	0.0086	0.0679	17
4	0.0710	0.0692	4.9099	9
5	0.0863	0.0865	7.4680	6
6	0.0475	0.0469	2.2263	10
7	0.0368	0.0354	1.3031	12
8	0.0036	0.0045	0.0163	19
9	0.0817	0.0805	6.5745	8
10	0.0286	0.0287	0.8216	13
11	0.0094	0.0100	0.0935	16
12	0.0400	0.0408	1.6320	11
13	0.0036	0.0045	0.0163	19
14	0.0072	0.0079	0.0574	18
15	0.0910	0.0916	8.3359	1
16	0.0832	0.0830	6.9041	7
17	0.0906	0.0913	8.2754	2
18	0.0898	0.0904	8.1125	4
19	0.0911	0.0906	8.2567	3
20	0.0894	0.0894	7.9881	5

Nowadays, the establishment of probabilistic multi-objective optimization methodology provides necessary conditions for multi-objective reliability optimization [1, 2]. In this section, the reliability design and assessment involving optimization with multiple objectives by means of probabilistic multi-objective optimization is developed.

In the reliability design and assessment involving optimization with multiple objectives by means of probabilistic multi-objective optimization, three cases are involved in general, i.e., first the reliability assessment is taken as additional restraint condition, second the reliability assessment is taken as an additional objective to be optimized, and in the last case the reliability assessment plays parts in both additional restraint condition and objective to be optimized.

In the first case, the optimization with multiple objectives by means of probabilistic multi-objective optimization can be done as usual with the additional restraint condition of reliability requirement; in the second case, the optimization with multiple objectives by means of probabilistic multi-objective optimization can be performed with the reliability assessment as an additional optimized objective to be conducted; in the third case, the mixture of above two cases is to be dealt with.

In the probabilistic multi-objective optimization, the objectives (responses) are divided into two types to be evaluated, respectively, i.e., the beneficial type of objective response and the unbeneficial type of objective response basically according to their roles or preferences.

In the following, we will develop the appropriate approach for reliability assessment involving optimization with multiple objectives by means of probabilistic multi-objective optimization by using an example to demonstrate the procedure.

The example belongs to the third case, i.e., take an example that involves both reliability and cost as mutual restraint conditions and as objectives to be optimized.

An example: design of a shaft with reliability and manufacturing cost as optimal objectives.

It is given that the strength and stress of a shaft obey normal distribution. The functional relationship between total manufacturing cost C and strength mean value μ_r, strength standard deviation σ_r, stress mean value μ_s and stress standard deviation σ_s can be approximately expressed as [40],

$$C(\mu_r, \sigma_r, \mu_s, \sigma_s) = C_1(\mu_r) + C_2(\sigma_r) + C_3(\mu_s) + C_4(\sigma_s). \tag{6.15}$$

According to the statistical data, the corrected average cost function related to strength, standard deviation of strength, stress and standard deviation of stress is obtained as follows [40],

$$C_1(\mu_r) = 0.2(\mu_r/10)^{1.5}, \tag{6.16}$$

$$C_2(\sigma_r) = 100(\sigma_r/10)^{-1.2}, \tag{6.17}$$

$$C_3(\mu_s) = 100(\mu_s/10)^{-0.6}, \tag{6.18}$$

$$C_4(\sigma_s) = 50(\sigma_s/10)^{-0.7}. \tag{6.19}$$

In Eqs. (6.16)–(6.19), the unit of stress and strength is MPa; the unit of costs is hundred yuan RMB. The allowable range of each distribution parameter is: $150 \leq \mu_r \leq 500$ MPa; $10 \leq \sigma_r \leq 80$ MPa; $50 \leq \mu_s \leq 250$ MPa; $10 \leq \sigma_s \leq 80$ MPa. It is required to ensure that the reliability of the parts is not less than 0.90 and the total manufacturing cost of the parts is not more than 120 hundred yuan RMB. The problem is to determine the values of each distribution parameter to make the reliability of the parts as high as possible and the total manufacturing cost as low as possible.

If both strength and stress follow normal distribution, the reliability function reads,

$$R(\mu_r, \sigma_r, \mu_s, \sigma_s) = \int_{-\infty}^{(\mu_r-\mu_s)/(\sigma_r^2+\sigma_s^2)^{0.5}} \left[\frac{1}{\sqrt{2\pi}} \cdot \exp\left(-\frac{1}{2}z^2\right)\right] \cdot dz. \tag{6.20}$$

The total manufacturing cost function is,

$$C(\mu_r, \sigma_r, \mu_s, \sigma_s) = 0.2(\mu_r/10)^{1.5} + 100(\sigma_r/10)^{-1.2} + 100(\mu_s/10)^{-0.6} + 50(\sigma_s/10)^{-0.7}. \tag{6.21}$$

Therefore, the optimization model of maximum reliability and minimum total cost under the conditions of total cost restraint and reliability restraint is,

$$\left.\begin{array}{l} \max R(\mu_r, \sigma_r, \mu_s, \sigma_s) \\ \min C(\mu_r, \sigma_r, \mu_s, \sigma_s) \\ s.t.: \\ R(\mu_r, \sigma_r, \mu_s, \sigma_s) \geq 0.90 \\ C(\mu_r, \sigma_r, \mu_s, \sigma_s) \leq 120 \\ 150 \leq \mu_r \leq 500 \\ 10 \leq \sigma_r \leq 80 \\ 50 \leq \mu_s \leq 250 \\ 10 \leq \sigma_s \leq 80 \end{array}\right\} \tag{6.22}$$

Quasi-uniform distribution $U_{53}(53^4)$ is used to spread the sampling points and conduct the test, and the results are shown in Table 6.22. There are 24 sampling points falling within the effective domain that satisfies the restraint conditions. The 24 effective sampling points can meet the basic needs of number of uniform sampling points in the sampling approach [1, 2]. In Table 6.22, z_c is an instantaneous parameter at each sampling point, $z_c = (\mu_r - \mu_s)/({\sigma_r}^2 + {\sigma_s}^2)^{0.5}$, and the symbol "$\surd$" indicates sampling point falling into the effective domain. Table 6.23 shows the evaluation results by means of probabilistic multi-objective optimization. In the evaluation, the reliability is a beneficial type of response and the total manufacturing cost is a unbeneficial type of response. The results show that the sampling point No. 8 has the highest total preferable probability and can be used as the optimum point for the system. Its total manufacturing cost is 75.4482 hundred yuan RMB, and its reliability is 92.36%. The corresponding parameter values of μ_r, σ_r, μ_s and σ_s are 199.5283 MPa, 46.3208 MPa, 108.4906 MPa and 43.6793 MPa, respectively. While the sampling point No. 2 followed it closely, its total manufacturing cost was 79.5063 hundred yuan RMB with reliability of 93.19%, and the corresponding parameter values of μ_r, σ_r, μ_s and σ_s are 159.9057 MPa, 54.2453 MPa, 63.2076 MPa and 35.7547 MPa, respectively.

Table 6.22 Distribution of sampling points with quasi-uniform distribution $U_{53}(53^4)$

No	x_1	x_2	x_3	x_4	μ_r	σ_r	μ_s	σ_s	120−C	z_c− 1.29	Validity
1	1	17	29	37	153.3019	31.7925	157.5472	58.2076	49.3435	−1.3540	
2	2	34	4	20	159.9057	54.2453	63.2076	35.7547	40.4937	0.1984	√
3	3	51	33	3	166.5094	76.6981	172.6415	13.3019	38.6870	−1.3688	
4	4	14	8	40	173.1132	27.8302	78.3019	62.1698	33.3100	0.1019	√
5	5	31	37	23	179.7170	50.2830	187.7358	39.7170	54.1103	−1.4151	
6	6	48	12	6	186.3208	72.7359	93.3962	17.2642	34.3834	−0.0470	
7	7	11	41	43	192.9245	23.8679	202.8302	66.1321	38.0873	−1.4309	
8	8	28	16	26	199.5283	46.3208	108.4906	43.6793	44.5519	0.1399	√
9	9	45	45	9	206.1321	68.7736	217.9245	21.2264	46.1317	−1.4538	
10	10	8	20	46	212.7358	19.9057	123.5849	70.0943	21.6853	−0.0665	
11	11	25	49	29	219.3396	42.3585	233.0189	47.6415	49.8828	−1.5046	
12	12	42	24	12	225.9434	64.8113	138.6792	25.1887	41.0694	−0.035	
13	13	5	53	49	232.5472	15.9434	248.1132	74.0566	13.5645	−1.4955	
14	14	22	28	32	239.1509	38.3962	153.7736	51.6038	41.4541	0.0374	√
15	15	39	3	15	245.7547	60.8491	59.4340	29.1509	26.2156	1.4715	√
16	16	2	32	52	252.3585	11.9811	168.8679	78.0189	−16.0683	−0.2323	
17	17	19	7	35	258.9623	34.4340	74.5283	55.5660	25.9483	1.5314	√
18	18	36	36	18	265.5660	56.8868	183.9623	33.1132	41.1627	−0.0502	
19	19	53	11	1	272.1698	79.3396	89.6226	10.6604	8.6357	0.9903	√
20	20	16	40	38	278.7736	30.4717	199.0566	59.5283	33.3372	−0.098	
21	21	33	15	21	285.3774	52.9245	104.7170	37.0755	31.5557	1.5058	√
22	22	50	44	4	291.9811	75.3774	214.1509	14.6226	25.3594	−0.2764	
23	23	13	19	41	298.5849	26.5094	119.8113	63.4906	20.0807	1.3084	√
24	24	30	48	24	305.1887	48.9623	229.2453	41.0377	37.5362	−0.1013	
25	25	47	23	7	311.7925	71.4151	134.9057	18.5849	22.3401	1.1070	√
26	26	10	52	44	318.3962	22.5472	244.3396	67.4528	18.5341	−0.2487	
27	27	27	27	27	325.0000	45.0000	150.0000	45.0000	29.3536	1.4599	√
28	28	44	2	10	331.6038	67.4528	55.6604	22.5472	7.6871	2.5899	√
29	29	7	31	47	338.2075	18.5849	165.0943	71.4151	1.9073	1.0559	√
30	30	24	6	30	344.8113	41.0377	70.7547	48.9623	13.7723	2.9998	√
31	31	41	35	13	351.4151	63.4906	180.1887	26.5094	24.5413	1.1987	√
32	32	4	10	50	358.0189	14.6226	85.8491	75.3774	−25.9124	2.2547	
33	33	21	39	33	364.6226	37.0755	195.2830	52.9245	22.8259	1.3306	√
34	34	38	14	16	371.2264	59.5283	100.9434	30.4717	15.1064	2.7518	√
35	35	1	43	53	377.8302	10.6604	210.3774	79.3396	−46.8697	0.8018	
36	36	18	18	36	384.4340	33.1132	116.0377	56.8868	10.7787	2.7876	√
37	37	35	47	19	391.0377	55.5660	225.4717	34.4340	21.8596	1.2427	√

(continued)

Table 6.22 (continued)

No	x_1	x_2	x_3	x_4	μ_r	σ_r	μ_s	σ_s	120−C	z_c− 1.29	Validity
38	38	52	22	2	397.6415	78.0189	131.1321	11.9811	−4.0550	2.0864	
39	39	15	51	39	404.2453	29.1509	240.566	60.8491	11.9408	1.1359	√
40	40	32	26	22	410.8491	51.6038	146.2264	38.3962	13.8796	2.8241	√
41	41	49	1	5	417.4528	74.0566	51.8868	15.9434	−16.2988	3.5357	
42	42	12	30	42	424.0566	25.1887	161.3208	64.8113	−0.6000	2.4885	
43	43	29	5	25	430.6604	47.6415	66.9811	42.3585	−2.0334	4.4148	
44	44	46	34	8	437.2642	70.0943	176.4151	19.9057	3.7577	2.2898	√
45	45	9	9	45	443.8679	21.2264	82.0755	68.7736	−20.9159	3.7367	
46	46	26	38	28	450.4717	43.6793	191.5094	46.3208	8.3768	2.7774	√
47	47	43	13	11	457.0755	66.1321	97.1698	23.8679	−4.9178	3.8290	
48	48	6	42	48	463.6792	17.2642	206.6038	72.7359	−23.7974	2.1488	
49	49	23	17	31	470.2830	39.7170	112.2642	50.2830	−3.1871	4.2974	
50	50	40	46	14	476.8868	62.1698	221.6981	27.8302	2.9716	2.4565	√
51	51	3	21	51	483.4906	13.3019	127.3585	76.6981	−51.983	3.2850	
52	52	20	50	34	490.0943	35.7547	236.7925	54.2453	−0.5802	2.6088	
53	53	37	25	17	496.6981	58.2076	142.4528	31.7925	−4.6550	4.0511	

6.9 Discussion on Integral Approach and Its Applications

The core requirements of systems theory are both the organic integrity of a system, and the optimization of a system being its overall optimization. Besides, the insepa-raty and irreplacement of every component in a system are the fundamental features of a system. It is no doubt that the issue of multi-objective optimization possesses all these characteristics.

The probabilistic multi-objective optimization of a system is based upon probability theory and set theory in viewpoint of systems theory, which reveals the simultaneity of optimization of multiple objectives; the treatments for both beneficial and unbeneficial types of utility indexes of performance indicators are equivalent and conformable.

Disturbances and changes of environmental factors lead to uncertainty of product quality. The robust design of production can be seen as a specific case of multi-objective optimization, both the mean value and the variation of objective responses are need to be dealt with in separate manner. Additionally, the reliability design and assessment involving optimization with multiple objectives is taken as another specific case of probabilistic multi-objective optimization as well.

Jaynes stressed the significance of making sure the meanings of "optimality" and robustness for a system so that important and proper information could be abstracted through the optimization process [41].

Table 6.23 Evaluation results by means of of probabilistic multi-objective optimization with quasi-uniform distribution $U_{53}(53^4)$

No	C, h RMB	z_c	R	P_c	P_R	$P_t \times 10^3$	Rank
2	79.5063	1.4884	0.9319	0.0507	0.0395	2.0017	2
4	86.6900	1.3919	0.9177	0.0475	0.0389	1.8471	8
8	75.4482	1.4299	0.9236	0.0525	0.0391	2.0545	1
14	78.5459	1.3274	0.9082	0.0511	0.0385	1.9673	4
15	93.7844	2.7615	0.9971	0.0444	0.0422	1.8736	6
17	94.0517	2.8214	0.9976	0.0443	0.0422	1.8695	7
19	111.3643	2.2803	0.9887	0.0366	0.0419	1.5304	20
21	88.4443	2.7958	0.9974	0.0467	0.0422	1.9745	3
23	99.9194	2.5984	0.9953	0.0416	0.0423	1.7552	13
25	97.6599	2.3970	0.9918	0.0426	0.0420	1.7913	11
27	90.6464	2.7499	0.9970	0.0458	0.0422	1.9324	5
28	112.3129	3.8799	0.9999	0.0361	0.0423	1.5299	21
29	118.0927	2.3459	0.9906	0.0336	0.0420	1.4078	24
30	106.2277	4.2898	0.99999	0.0388	0.0424	1.6447	16
31	95.4587	2.4887	0.9936	0.0436	0.0421	1.8357	9
33	97.1741	2.6206	0.9956	0.0429	0.0422	1.8072	10
34	104.8936	4.0417	0.99,997	0.0394	0.0424	1.6698	14
36	109.2213	4.0776	0.99,998	0.0375	0.0424	1.5883	18
37	98.1404	2.5327	0.9942	0.0424	0.0421	1.7866	12
39	108.0592	2.4259	0.9925	0.0380	0.0420	1.5981	17
40	106.1204	4.1141	0.99,998	0.0389	0.0424	1.6467	15
44	116.2423	3.5798	0.9998	0.0344	0.0423	1.4558	22
46	111.6232	4.0674	0.99,998	0.0364	0.0424	1.5430	19
50	117.0284	3.7465	0.99,991	0.0340	0.0423	1.4411	23

The fuzzification approach of robust design for case of desired value as target is developed in terms of probabilistic multi–objective optimization in this section. The consequences of the robust designs of welding process parameters of GMA welding, parametric design of printing machine's ability and robust parametric design for coating thickness of bare silicon wafers quality controlling indicate the reasonability of the fuzzification approach as an auxiliary measure. Reliability design and assessment involving optimization with multiple objectives is also a specific case of probabilistic multi-objective optimization.

All application examples reveal the validity of the methodology of probabilistic multi–objective optimization for a system.

6.10 Concluding Remarks

From above description, the probabilistic multi-objective optimization from viewpoint of systems theory is given briefly, which reveals the simultaneity of optimization of multiple objectives and treats both beneficial type of utility indexes and unbeneficial type of utility indexes of performance indicators equivalently and conformably. Robust design could be seen as a special case of probabilistic multi-objective optimization. The fuzzification approach of robust design for case of desired value as target is an auxiliary and nimble measure for probabilistic multi–objective optimization. In addition, the reliability design and assessment involving optimization with multiple objectives is taken as another specific case of probabilistic multi-objective optimization as well. The approach is impersonal without any subjective or personal factors.

References

1. M. Zheng, J. Yu, H. Teng, Y. Cui, Y. Wang, *Probability-Based Multi-objective Optimization*, 2nd edn. (Springer, Singapore, 2023)
2. M. Zheng, J. Yu, *Robust Design and Assessment of Product and Production by Means of Probabilistic Multi-objective Optimization* (Springer, Singapore, 2024)
3. S. Opricovic, G.H. Tzeng, Compromise solution by MCDM methods: a comparative analysis of VIKOR and TOPSIS. Eur. J. Oper. Res. **156**, 445–455 (2004)
4. A. Shanian, O. Savadogo, Multiple–criteria decision support analysis for material selection of metallic dipolar plate for polymer electrolyte fuel cell. J. Power Sources **159**, 1095–1104 (2006)
5. M.B. Babanli, F. Prima, P. Vermaut, L.D. Demchenko, A.N. Titenko, S.S. Huseynov, R.J. Hajiyev, V.M. Huseynov, Material selection methods: a review, *in Advances in Intelligent Systems and Computing 896, 13th International Conference on Theory and Application of Fuzzy Systems and Soft Computing–ICAFS–2018*, ed. by R.A. Aliev, J. Kacprzyk, W. Pedrycz, M. Jamshidi, F. M. Sadikoglu (Springer, Cham, 2019), pp. 929–936. https://doi.org/10.1007/978-3-030-04164-9_123
6. B.M. Ayyub, R.H. McCuen, *Probability, Statistics, and Reliability for Engineers and Scientists*, 3rd edn. (CRC Press, Taylor & Francis Group, A Chapman & Hall Book, 2011, Boca Raton). 978-1-4398-9533-7. (eBook-PDF)
7. W. Yang, S. Chon, C. Choe, J. Yang, Materials selection method using TOPSIS with some popular normalization methods. Eng. Res. Express **3**, 015020 (2021)
8. V. Modanloo, A. Doniavi, R. Hasanzadeh, Application of multi criteria decision making methods to select sheet hydroforming process parameters. Decis. Sci. Lett. **5**(3), 349–360 (2016)
9. M. Moradian, V. Modanloo, S. Aghaiee, Comparative analysis of multi criteria decision making techniques for material selection of brake booster valve body. J. Traffic. Trans. Eng. **6**, 526–534 (2019). https://doi.org/10.1016/j.jtte.2018.02.001
10. V. Modanloo, V. Alimirzaloo, M. Elyasi, Multi–objective optimization of the stamping of Titanium bipolar plates for fuel cell. Int. J. Adv. Des. Manuf. Technol. **12**(4), 1–8 (2019)
11. I.Y. Kim, O. de Weck, Adaptive weighted sum method for multiobjective optimization: a new method for Pareto front generation. Struct. Multidiscip. Opt. **31**(2), 105–116 (2006)
12. J. Ye, *System Science* (Sichuan Academy of Social Sciences Press, Chengdu, 1987)
13. G.E. Mobus, *Systems Science: Theory, Analysis, Modeling, and Design* (Springer, Cham, 2022)

14. Y.S. Vasiliev, N.D. Pankratova, V.N. Volkova, O.D. Shipunova, N.N. Lyabakh, *System Analysis in Engineering and Control* (Springer, Cham, 2022)
15. N. Rezaei, *Integrated Science, Science Without Borders* (Springer, Cham, 2021)
16. D. Wang, *Probability Theory and Mathematical Statistics* (Beijing Institute of Technology Press, Beijing, 2020)
17. M. Zheng, J. Yu, Rational option of optimum parameters of robust design of hybrid of PMOO with sequential uniform algorithm. J. Umm Al-Qura Univ. Eng. Archit. (2024). https://doi.org/10.1007/s43995-024-00058-0
18. X. Zhang, J. Chen, H. Shi, *Mechanical Optimization Design*, 2nd edn. (Science Press, Beijing, 2023)
19. G.E.P. Box, N.R. Draper, *Empirical Model Building and Response Surfaces* (Wiley, New York, 1987)
20. G. Taguchi, *The System of Experimental Design Engineering Methods to Optimize Quality and Minimize Cost* (Productivity Press, New York, 1987)
21. V.N. Nair, Taguchi's parameter design: a panel discussion. Technometrics **34**(2), 127–161 (1992)
22. G.G. Vining, R.H. Myers, Combining Taguchi and response surface philosophies: a dual response approach. J. Qual. Technol. **22**(1), 38–45 (1990)
23. E.D. Castillo, D.C. Montgomery, A nonlinear programming solution to the dual response problem. J. Qual. Technol. **25**(3), 199–204 (1995)
24. D.K.J. Lin, W. Tu, Dual response surface optimization. J. Qual. Technol. **27**(1), 34–39 (1995)
25. A.F.C. Karen, P.R. Nelson, Dual response optimization via direct function minimization. J. Qual. Technol. **28**(3), 26–30 (1996)
26. K.J. Kim, D.K.J. Lin, Dual response surface optimization: a fuzzy modeling approach. J. Qual. Technol. **30**(1), 1–10 (1998)
27. R. Ding, K.J. Lin, D. Wei, Dual response surface optimization: a weighted MSE approach. Qual. Eng. **16**(3), 377–385 (2004)
28. J. Kovach, B.R. Cho, J. Antony, Development of a variance prioritized multi-response robust design framework for quality improvement. Int. J. Qual. Reliab. Manag. **26**(4), 380–396 (2009)
29. I.J. Jeong, K.J. Kim, D.K.J. Lin, Bayesian analysis for weighted mean squared error in dual response surface optimization. Qual. Reliab. Eng. Int. **26**(5), 417–430 (2010)
30. D.H. Lee, I.J. Jeong, K.J. Kim, A posterior preference articulation approach to dual-response-surface optimization. IIE Trans. **42**(2), 161–171 (2010)
31. Z. He, Y. H. Ma, Y. Zhao, Multi-response robust optimization design based on Taguchi process capability index and entropy weight theory. Chin. Agric. Mech. **3**, 33–36 (2008). 1006-7205(2008)03-0033-04
32. L. Ouyang, Y. Ma, J. Wang, F. Wu, Robust design based on entropy weight and dual response surface. J. Ind. Eng. Eng. Manag. **28**(2), 191–195 (2014). https://doi.org/10.13587/j.cnki.jieem.2014.02.015
33. M. Zheng, J. Yu, Probabilistic approach for robust design with orthogonal experimental methodology in case of target the best. J. Umm Al-Qura Univ. Eng. Archit. **15**, 55–59 (2024). https://doi.org/10.1007/s43995-023-00040-2
34. F. Yang, T. Wang, S. Yang, Minimization of warpage of plastic inj ection molded parts of ABS by Taguchi method. J. Plast. Eng. **15**(5), 195–197 (2008)
35. D.M. Byrne, S. Taguchi, The Taguchi approach to parameter design. Qual. Prog. **20**(12), 19–26 (1987)
36. D. Kim, S. Rhee, Optimization of GMA welding process using the dual response approach. Int. J. Prod. Res. **41**(18), 4505–4515 (2003). https://doi.org/10.1080/0020754031000/595800
37. S.M. Pickle, T.J. Robinson, J.B. Birch, C.M. Anderson-Cook, A semi-parametric approach to robust parameter design. J. Stat. Plan. Inference **138**, 114–131 (2008)
38. G.G. Vining, R.H. Myers, Combining Taguchi and response surface philosophies: a dual response approach. J. Quality Technol. **22**, 38–45 (1990)
39. S. Shin, B.R. Cho, Robust design models for customer-specified bound on process parameters. J. Syst. Sci. Syst. Eng. **15**(1), 2–18 (2006)

40. G. Bai, R. Wang, Satisfactory solution of multi-objective optimization design of mechanical parts reliability. China Mech. Eng. **79**(8), 851–853 (2001)
41. E.T. Jaynes, *Probability Theory: The Logic of Science* (Cambridge University Press, Cambridge, 2003)

Chapter 7
Concise Description of Effective Media Approximation

Abstract This chapter gives a concise description of effective media approximation, which considers the homogeneity and ergodicity of constituting components of a composite system in macroscale grossly in statistical physics. The effective elasticity modulus of grains reinforced composite materials, strength and toughness parameters of porous materials, effective thermal expansion coefficient, effective thermal conductivity, effective dielectric constant, effective magnetic permeability, and effective electrical conductivity of randomly distributed phases composite, etc., are concerned.

Keywords System · Composite material · Statistical homogeneity · Ergodicity · Effective medium approximation · Macroscale · Detectable · Resolution

7.1 Introduction

The *Effective-Medium Approximation* (EMA) is a very convenient and useful tool to describe the material response of inhomogeneous systems, of which the non-uniformity is in smaller size (microsize) than the detectable resolution (macroscale) [1]. The non-uniformity of constituting parts and the scattering actions will be only considered in cases of the detectable resolution (size) being comparable with or smaller than the size of non-uniformity.

Actually, the main role of EMA is to describe some "averaged response" of a system as a whole in macroscale by taking into account the distribution of size, shape and the response of constituting parts as well as their mutual interactions though with micro-non-uniformity. The main approaches include the Lorentz cavity idea for the determination of the polarizability of an individual inclusion and Clausius–Mossotti relation [1]. Bortchagovsky et al. realized that the two-dimensional (2D) situation is principally different. Inclusions or particles are spreading only in one plane, and the average near-field scattered by closest neighbors is not zero in this case and both near and far scattered fields contribute to the additional polarization. Besides, interactions of dipoles in both directions of along and across the plane are different which produce

M. Zheng and J. Yu, *Systems Theory for Engineering Practice*,
https://doi.org/10.1007/978-981-97-9342-6_7

geometrical anisotropy in the optical properties of such a layer. They developed the model based on the solution of the Lipman-Schwinger equation for averaged fields and polarizability, which aimed to describe the response for the external field as the 2D analogue of EMA [1].

In principle, EMA is based on the hypothesis of homogeneity and ergodicity of constituting parts in macro-scale in statistical physics; each constituting part immerses in the hypothetical background sea of the "effective media" constituting of components of the composite system, and volume percentage of each part takes part in the EMA as the appearing probability to provide a series of event; furthermore, specific equilibrium conditions are fulfilled concisely due to the fundamental requirements in physics, which thus leads to meaningful approaches for corresponding cases, i.e., the so called Effective-Medium Approximation.

Here in this chapter, typical approaches considering statistical homogeneity and ergodicity of constituting parts in macroscale are concisely described by using specific balance requirements in physics.

7.2 Effective Elasticity Modulus of Grains Reinforced Composite Materials

1. ***2-dimensional problems***

Considering that a round plate with radius r_I is embedded into a hole of radius r_2 in an infinite sheet in two-dimensional (2-d) space. If $r_1 > r_2$, it is called an overmatching problem in 2-d space [2]. The difference of r_1 and r_2 is defined as the overmatching value δ, i.e., $\delta = r_1 - r_2$.

In general, δ is much smaller than both r_I and r_2 such that the deformation due to this overmatching keeps in elastic range. Furthermore, let the Young's modulus and Poisson's ratio of the round plate and the infinite sheet be E_1 and ν_1, E_2 and ν_2, respectively, and the final equilibrium radius of the hole be r_0 due to the overmatching from r_2. Thus it appears a contact pressure, *P*, on the interface of the round plate and the infinite sheet due to overmatching process. Mechanically, the contact pressure *P* can be written as [2],

$$P = \frac{\delta}{r_0} \cdot \frac{1}{\frac{1-\nu_1}{E_1} + \frac{1+\nu_2}{E_2}}. \tag{7.1}$$

In Eq. (7.1), r_0 is determined by Eq. (7.2) [2],

$$r_0 = r_2 \cdot \left(1 + \frac{(1+\nu_2) \cdot E_1}{(1+\nu_2) \cdot E_1 + (1-\nu_1) \cdot E_2} \cdot \frac{\delta}{r_2}\right). \tag{7.2}$$

It is well known that order of magnitude of δ/r_2 is at *P/E*, i.e., $\delta/r_2 \sim P/E$, in the range of elasticity, and that *P* is always less than the elastic limit, P_e. In general, the value of P_e/E is in the range of 10^{-4}–10^{-2} for metals [2]. Therefore, Eq. (7.2) can be approximately simplified by

$$r_0 \approx r_2. \tag{7.3}$$

As a result, Eq. (7.1) can be approximately written as,

$$P \approx \frac{\delta}{r_2} \cdot \frac{1}{\frac{1-\nu_1}{E_1} + \frac{1+\nu_2}{E_2}} \tag{7.4}$$

Obviously, Eq. (7.4) is with very high accuracy though it is an approximation.

As to an ordinary situation when the material of the round plate is the same as that of the infinite sheet, then, $E_1 = E_2 \equiv E_0$ and $\nu_1 = \nu_2 \equiv \nu_0$, it yields,

$$P \approx \frac{\delta}{r_2} \cdot \frac{E_0}{2}. \tag{7.5}$$

Now, let's study a grain-reinforced composite, which is homogenous and isotropic in macro-scale with an effective modulus, E_0, and effective Poisson's ratio, ν_0. Furthermore, the granular composite is taken to make the round plate and the infinite sheet for overmatching each other, as mentioned above. Therefore, the pressure, P, on the interface of the round plate and the infinite sheet can be expressed by Eq. (7.5).

On the other hand, the grain-reinforced composite consists of n phases in micro-scale. The volume fraction of the i-th phase in the grain-reinforced composite is x_i, and the elasticity modulus and Poisson's ratio of the i-th phase are E_i and ν_i, respectively. Based on the ergodic hypothesis of statistical physics, the grain- reinforced composite disc may be reasonably taken as the equivalence of an ideal round plate, which is a pileup of n discs with the same radius, r_2, and the appearance probabilities of the n discs in the pile are $x_1, x_2, \ldots x_i, \ldots$, and x_n, respectively. The Young's modulus and Poisson's ratio of the n discs are $E_1, E_2, \ldots E_i, \ldots$, and E_n, and $\nu_1, \nu_2, \ldots \nu_i, \ldots$, and ν_n. In condition of the i-th disc being only taken to make overmatching with the infinite sheet of grains reinforced composite mentioned above, the pressure (called partial pressure), P_i, on the interface of the i-th disc and the grain- reinforced composite sheet mentioned above can be approximately written as,

$$P_i = \frac{\delta}{r_2} \cdot \frac{1}{\frac{1-\nu_i}{E_i} + \frac{1+\nu_0}{E_0}}. \tag{7.6}$$

As a result, the actual pressure, P, on the interface of the grain-reinforced composite plate and the infinite sheet with the same grain-reinforced composite materials due to the overmatching, is the summation of the n partial pressure in the first-order approximation, i.e.,

$$P = \sum_{i=1}^{n} x_i \cdot P_i = \sum_{i=1}^{n} x_i \cdot \frac{\delta}{r_2} \cdot \frac{1}{\frac{1-\nu_i}{E_i} + \frac{1+\nu_0}{E_0}}. \tag{7.7}$$

Alternatively the actual pressure, *P*, on the interface of the grain-reinforced composite plate and the infinite sheet with the same grain-reinforced composite material due to overmatching is Eq. (7.5), which thus leads to following requirement,

$$\sum_{i=1}^{n} x_i \cdot \frac{1}{\frac{1-\nu_i}{E_i} + \frac{1+\nu_0}{E_0}} = \frac{E_0}{2}. \tag{7.8}$$

An alternative form of Eq. (7.8) is,

$$\sum_{i=1}^{n} \frac{2x_i \cdot E_i}{(1-\nu_i) \cdot E_0 + (1+\nu_0) \cdot E_i} = 1. \tag{7.9}$$

Equation (7.9) is in fact the expression for predicting the elasticity modulus of 2-d grain-reinforced composite.

An application example was the comparison of predictions of elasticity modulus of the grain-reinforced composite including two phases by using Eq. (7.9) with those of the simulation with computational micromechanics [2], whose elasticity moduli were taken as $E_l = 70{,}000$ MPa and $E_2 = 200{,}000$ MPa. The comparison of predictions of elasticity modulus of above granular composite by using Eq. (7.9) with those of the computational micromechanics is shown in Fig. 7.1, in the estimations of prediction, $\nu_1 = \nu_2 \approx 0.3$ is employed for metallic materials. It can be seen from Fig. 7.1 that the values of the elasticity modulus predicted by using Eq. (7.9) agree with those of the simulation [2].

As to a porous material, voids appears in materials randomly, which can be taken as the second phase with the elasticity modulus $E_2 = 0$, then the expression for predicting the effective elasticity modulus of the materials containing randomly distributed voids can be obtained approximately by substituting $E_2 = 0$ into Eq. (7.9),

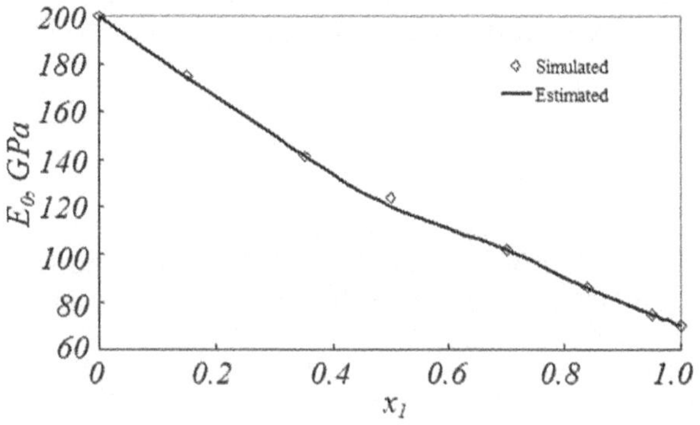

Fig. 7.1 Comparison of predictions of elasticity modulus of composite by using Eq. (7.9) with simulation for 2-d

it yields,

$$E_0 = E_1 \cdot [1 - 2.86(1 - x_1)] = E_1 \cdot (1 - 2.86\varphi). \tag{7.10}$$

In Eq. (7.10), E_0 is the effective elasticity modulus of the porous material containing random voids; φ is the volume fraction of the voids, $\varphi = 1 - x_1$.

The comparison of Eq. (7.10) with the test data is shown in Ref. [2], which showed good agreement of predicted E_0 by Eq. (7.10) with the test results in range of $\varphi < 0.25$.

Besides, Eq. (7.10) shows that the value of E_0 approaches to zero when the volume fraction φ of the voids reaches to its critical value $\varphi_{c2} = 0.35$. This phenomenon is called the elastic percolation failure [2].

2. ***3-dimensional problems***

Analogically, a model to deal with a 3-dimensional (3-d) problem was also proposed [2]. It considers that a sphere of radius r_l is embedded into the spherical cavity of radius r_2 in an infinite medium in 3-d space. Again, if $r_1 > r_2$, this is an overmatching problem in 3-d space. The difference between r_l and r_2 is δ, and there is $\delta \ll r_1, r_2$.

Similarly, with the analogical derivations, it yields an expression of effective elasticity modulus for 3-d grain-reinforced composite [2],

$$\sum_{i=1}^{n} \frac{x_i}{\frac{(1-2\nu_i)}{E_i} + \frac{(1+\nu_0)}{2E_0}} = \frac{2E_0}{3(1-\nu_0)}. \tag{7.11}$$

Besides, the comparison of Eq. (7.11) with the tests is redrawn in Fig. 7.2 [2], which shows a good agreement with the test results.

Similarily, as to a porous material in 3-d, the elasticity modulus of voids can be set as $E_2 = 0$, the expressions for predicting the effective elasticity modulus of the porous

Fig. 7.2 Comparison of Eq. (7.11) with the tests for 3-d

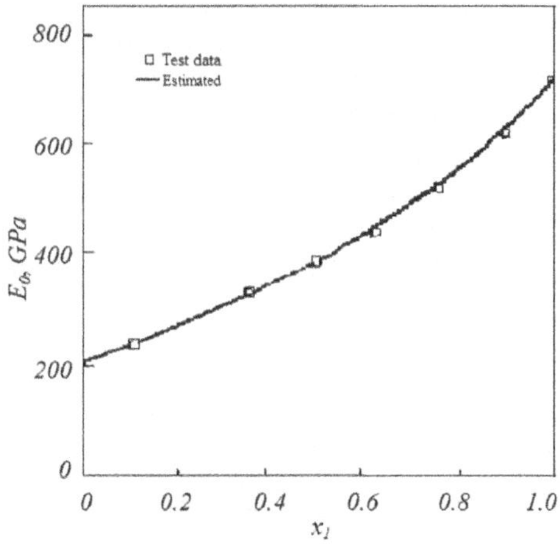

materials containing randomly distributed voids can be obtained by substituting $E_2 = 0$ into Eq. (7.11), it yields,

$$E_0 = E_1 \cdot [1 - 2.625(1 - x_1)] = E_1 \cdot (1 - 2.625\varphi). \tag{7.12}$$

Obviously, the value of E_0 approaches to zero when the volume fraction φ of the voids reaches to its critical value $\varphi_{c3} = 0.381$ for 3-d porous materials [2].

7.3 Strength and Toughness Parameters of Porous Materials

Furthermore, the strength and toughness parameters of porous materials were obtained [3–5]. The yielding strength of porous materials was derived in the viewpoint of distortional energy explanation for von Mises yielding critieron [3], which is,

$$\sigma_y^2 + 6.25 \cdot \varphi \cdot \sigma_m^2 = (1 - \varphi) \cdot (1 - 2.625\varphi)\sigma_s^2, \tag{7.13}$$

where, σ_y and σ_m express apparent equivalent stress and mean stress (hydrostatic stress) of the porous material, respectively; φ is the volume percentage of voids; σ_s indicates the yielding strength of matrix (voids free) material.

For uni-axial tension and compression, $\sigma_1 = \sigma_y$, $\sigma_2 = \sigma_3 = 0$, $\sigma_m = \sigma_1/3$, it yields [3],

$$\sigma_1^2 + \frac{6.25 \cdot \varphi}{9} \cdot \sigma_1^2 = (1 - \varphi) \cdot (1 - 2.625\varphi)\sigma_s^2, \tag{7.14}$$

$$\sigma_1 = \sqrt{\frac{(1 - \varphi) \cdot (1 - 2.625\varphi)}{\left(1 + \frac{6.25\varphi}{9}\right)}} \cdot \sigma_s. \tag{7.15}$$

From Eqs. (7.14) and (7.15), the apparent critical yielding strength of porous maietrals in uni-axial tension and compression can be determined. A comparison of experimental data of apparent yielding strength for porous matrial is redrawn in Fig. 7.3 [3].

Furthermore, if the maitrx of the work-hadrening porous mateirals obeys the role of Eq. (7.16),

$$\sigma_e = K_0 \cdot \varepsilon^n, \tag{7.16}$$

where σ_e is the flow stress of matirx; ε is the equivalent strain; n is the work-hardening exponent; K_0 is the strength coeiffcient of matrix. Then it yields,

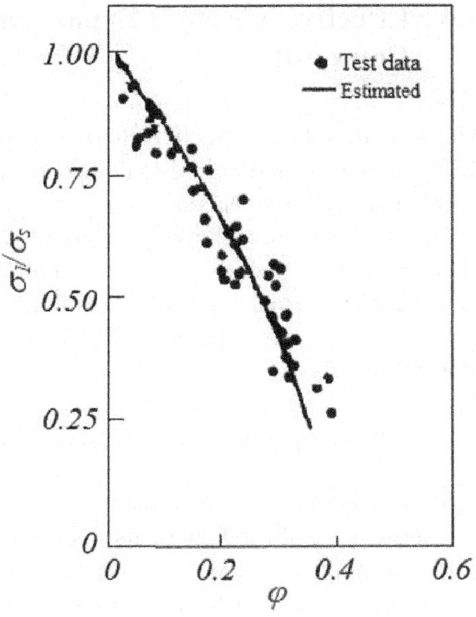

Fig. 7.3 Comparison of experimental data of apparent yielding strength for porous matrial

$$K = \sqrt{\frac{(1-\varphi)\cdot(1-2.625\varphi)}{\left(1+\frac{6.25\varphi}{9}\right)}}\cdot K_0. \tag{7.17}$$

Equation (7.17) is the expression of relationship among apparent strength coeiffcient K, strength coeiffcient of matrix K_0 and volume fraction φ.

In addition, the apparent fracture toughness K_{IC} of porous materials can be written as [4],

$$K_{IC}/K_{IC0} = \sqrt{(1-\varphi^{2/3})\cdot(1-2.625\varphi)}, \tag{7.18}$$

where K_{IC0} is fracture toughness of voids free (matrix) materials.

The apparent ultimate tensile strength S_b of porous materials is [4],

$$S_b/S_{b0} = \sqrt{\frac{(1-\varphi)\cdot(1-2.625\varphi)}{\left(1+\frac{6.25\varphi}{9}\right)}}, \tag{7.19}$$

where S_{b0} is ultimate tensile strength of voids free (matrix) materials.

7.4 Effective Thermal Expansion Coefficient of Random Composite

The new fomula of effective thermal expansion coefficient of a composite with different phases randomly distributed was presented [5], which took the equiliburm of thermal stress into account.

Assume a sphere with radius r, modulus E_{i}, thermal expansion coeiffeient α_{i}, and poisson ratio ν_{i}, is embedded into a hole with the same radius in a medium with modulus E_0, thermal expansion coemeient α_0, and Poisson ratio ν_0. If the interface is free of stress at temperature T_0, then, the thermal stress at the interface at any temperature T is

$$P_i = (\alpha_i - \alpha_0) \cdot (T - T_0)/[(1 + \nu_0)/(2E_0) + (1 - 2\nu_i)/E_i]. \tag{7.20}$$

If the sphere and the media are of the same material, then $P = 0$.

Consider the above sphere and its media to be both made of the same composite with n components randomly distributed, in macroscopic view, the interface stress P is zero. It is known from physics that P can be accurately approximated as the sum of partial pressures P_i induced by each component with its volume fraction x_i as follows,

$$P = \sum_{i=1}^{n} x_i \cdot P_i = 0. \tag{7.21}$$

The partial pressure P_i of i-th component can be expressed as Eq. [7.20], thus [5],

$$\alpha_0 = \left\{\sum_{i=1}^{n} x_i\alpha_i/[(1 + \nu_0)/(2E_0) + (1 - 2\nu_1)/E_1]\right\} \Big/ \left\{\sum_{i=1}^{n} x_i/[(1 + \nu_0)/(2E_0) + (1 - 2\nu_1)/E_1]\right\}. \tag{7.22}$$

The formulae for computing E_0 and ν_0, are rewritten as following, respectively [5],

$$\sum_{i=1}^{n} \frac{x_i}{\frac{(1-2\nu_i)}{E_i} + \frac{(1+\nu_0)}{2E_0}} = \frac{2E_0}{3(1 - \nu_0)}. \tag{7.23}$$

$$\nu_0 = \sum_{i=1}^{n} x_i \cdot \nu_i. \tag{7.24}$$

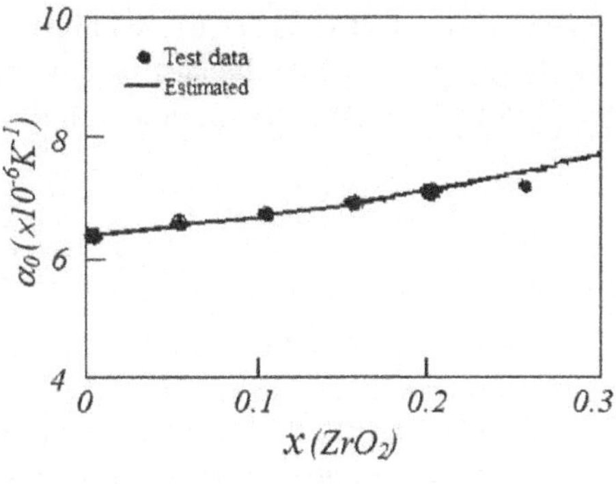

Fig. 7.4 Effective thermal expansion coefficient versus volume fraction of ceramic phase ZrO_2 in the doped low melting point glass (80% B_2O_3,10% BaO_3, 10%Al_2O_3)

Figure 7.4 is the redrawn of the relationship between the effective thermal expansion coefficient and the volume fraction of ceramic phase ZrO_2 in the doped system of low melting point glass (80% B_2O_3,10% BaO_3, 10%Al_2O_3) [5].

7.5 Effective Thermal Conductivity, Dielectric Constant, Magnetic Permeability, and Electrical Conductivity of Composite with Random Phases

Furthermore, effective medium theories for calculating the effective thermal conductivity, dielectric constant, electrical conductivity, magnetic permeability and other physical quantities of composites with different phases randomly distributed are derived according to the egodic hypothesis (or principle of equal probability) in statistical physics [6]. For the steady temperature field, the steady current field, the electrostatic field, and the potential of the static magnetic field in the medium, they all follow Laplace equation [6], i.e.,

$$\nabla^2 V = 0. \tag{7.25}$$

In Eq. (7.25), V represents temperature field, electric potential field, electric current field and magnetic scalar potential field for above four physical processes.

When the interface of medium is encountered on the spot, the following boundary value relationship is satisfied,

$$V_1 = V_2, \lambda_1 \cdot \left(\frac{\partial V}{\partial n}\right)_1 = \lambda_2 \cdot \left(\frac{\partial V}{\partial n}\right)_2. \tag{7.26}$$

In Eq. (7.26), the subscripts 1 and 2 represent different media on both sides of the interface; n represents the normal direction of the interface. λ expresses the thermal conductivity, electrical conductivity, dielectric constant and magnetic permeability for above four processes, respectively.

Since Eqs. (7.25) and (7.26) are universal for the above four processes, the discussion here is forcused on the electrostatic field only. It is known that the potential of the uniform external field E_0 in an infinite medium with a dielectric constant of ε_m is,

$$V = -\vec{E_0} \cdot \vec{R}. \tag{7.27}$$

In Eq. (7.27), $\vec{R}$ is the position vector of a point from a coordinate origin in space.

If a uniform dielectric sphere with a dielectric constant of ε_i is embedded in the infinite medium with dielectric constant of ε_m, and the center of the sphere is taken as the coordinate origin, then the space potential is [6],

$$V_{i1} = -\vec{E_0} \cdot \vec{R} + \frac{\varepsilon_i - \varepsilon_m}{\varepsilon_i + 2\varepsilon_m} \cdot \frac{R_0^3}{R^3} \cdot \vec{E_0} \cdot \vec{R}, \quad (R > R_0), \tag{7.28}$$

$$V_{i2} = -\vec{E_0} \cdot \vec{R} + \frac{\varepsilon_i - \varepsilon_m}{\varepsilon_i + 2\varepsilon_m} \cdot \vec{E_0} \cdot \vec{R}, \quad (R < R_0), \tag{7.29}$$

In Eqs. (7.28) and (7.29), R_0 is the radius of dielectric sphere.

Now, let's consider a multiphase dielectric medium with good randomness, which is used to make the above sphere and the outer infinite medium, thus Eq. (7.27) holds for it grossly since it is a homogeneous medium in macroscale. Denote x_i and ε_i represent the volume fraction and dielectric constant of the i-th component, then it derives following relationship according to egodic hypothesis in statistical physics [6],

$$\sum_{i=1}^{n} x_i \cdot \frac{\varepsilon_i - \varepsilon_m}{\varepsilon_i + 2\varepsilon_m} = 0. \tag{7.30}$$

Equation (7.30) determines the value of the effective dielectric constant ε_m of the multi phase composite from the individual dielectric constant ε_i and volume fraction x_i of its i-th components.

Landau once proposed an estimation for effective dielectric constant ε_m of the multi-phase composite [7], i.e.,

$$\varepsilon_m^{1/3} = \sum_{i=1}^{n} x_i \cdot \varepsilon_i^{1/3}. \tag{7.31}$$

The fundamental requirement for Eq. (7.31) valid is that the difference of dielectric constants ε_i among components is much smaller than the effective dielectric constant ε_m of the multi-phase composite [7].

Similarly, the relationship among effective thermal conductivity K_m of composite and its component thermal conductivity K_i and volume fraction x_i is [6],

$$\sum_{i=1}^{n} x_i \cdot \frac{K_i - K_m}{K_i + 2K_m} = 0. \tag{7.32}$$

Analogically, the relationship among gross effective electric conductivity σ_m of composite and its component electric conductivity σ_i and volume fraction x_i is [6],

$$\sum_{i=1}^{n} x_i \cdot \frac{\sigma_i - \sigma_m}{\sigma_i + 2\sigma_m} = 0. \tag{7.33}$$

While the relationship among gross effective magnetic conductivity μ_m of composite and its component magnetic permeability μ_i and volume fraction x_i is [6],

$$\sum_{i=1}^{n} x_i \cdot \frac{\mu_i - \mu_m}{\mu_i + 2\mu_m} = 0. \tag{7.34}$$

7.6 Analysis of Thermal Stress in Granular Composite

Subsequently, analysis of thermal stress in granular composite is presented [8].

A model of inletting reinforced grain ball and spherical hollow matrix "double embedding" into the effective medium of the granular composite was proposed [8]. The distributions of thermal stresses in different phases of the composite during cooling were investigated, a set of expressions for predicting thermal stresses in elastic and elastoplastic states were derived. It found that the reinforcement suffers compressive hydrostatic stress and the hydrostatic stress in matrix is a tensile one during cooling; as temperature continues to decrease, a yield zone will appear at the interface of reinforcing particle and the matrix, which extends outward from the interface; with the increase of the volume fraction of the reinforcing phase, the compressive stress of the reinforcing particles and the hydrostatic tensile stress in the matrix decrease, and the critical temperature difference of initial plastic yielding at the interface of reinforcing particle and the matrix increases, while the critical temperature difference of overall plastic yielding of the matrix material decreases [8].

7.7 Summary

From above description, it shows that effective media approximations are efficient approaches, which are based on the consideration of homogeneity and ergodicity of constituting parts in statistical physics in macro-scale; a scene of each constituting part being embedded in the background sea of the "effective media" constituting of components of the composite system one after another is displayed, of which the volume percentage is taken as the appearing probability; besides, specific balance requirements are concisely performed in physics, which thus derive meaningful approaches for corresponding cases. Coordinate collective actions and effects among phases are revealed in the specific equilibrium requirements of the derivations particularly.

References

1. E.G. Bortchagovsky, A. Dejneka, L. Jastrabik, V.Z. Lozovski, T.O Mishakova, Model of the effective-medium approximation for nanostructured layers with the account of interparticle interactions and its ellipsometric registration. Accad. Pelorit. Pericol. Cl. Sci. Fis. Mat. Nat. **89**(1), C1V89S1P019 (2011)
2. M. Zheng, X. Zheng, Expressions for predicting the elasticity modulus of materials reinforced by second-phase grains. Metal. Trans. **22A**, 507–511 (1991)
3. M. Zheng, Z. Luo, X. Zheng, The yielding behavior of materials with random voids. Chin. Sci. Bulletin **37**(6), 512–516 (1992)
4. M. Zheng, Z. Luo, X. Zheng, Intensity and toughness parameters of porous materials. Chin. Sci. Bull. **39**(10), 810–814 (1994)
5. M. Zheng, X. Zheng, A new formula of predicting effeetive thermal expansion coefficient of random composite. J. Northwest. Polytech. Univ. **12**(3), 499–501 (1994)
6. M. Zheng, Effective medium theory for calculating effective physical quantities of multiphase systems. J. Electron. Sci. Technol. **3**(2), 20–25 (1991)
7. L.D. Landau, E.M. Lifshitz, Continous medium electrodynamics, in *Course of Theoretical Physics*, vol. 08, 3rd edn. (Butterworth–Heinenann, Linacre House, Jordan Hill, UK, 2000)
8. M. Zheng, Z. Jin, H. Hao, A new model for analysing thermal stress in granular composite. Sci. China (series A) **38**(6), 739–748 (1995)

Chapter 8
Certainty in Uncertain Phenomena

Abstract The Galton knocked boards, light diffraction, electron diffraction, Benford's law, parallels between modern physics and Eastern mysticism, and Goodhart's law, are outlined in this chapter. It reveals that the overall certainty forms from the individual uncertain phenomenon, and the "unity of opposites" exists inside a system. The wholeness of things leads to the objective laws of the real world, which agrees with the global view of things of us obligatorily.

Keywords Certainty · Uncertain phenomena · Overall certainty · Unity · Opposites · Philosophy · Dialectics

8.1 Introduction

In natural, uncertain phenomena usually contain certainty. In the experiment of flipping a uniform coin, one could not know which side being face up in one individual flipping, but when the number of the flipping times is very large, the probabilities of each side being face up are nearly the same, i.e., with the probability of about 0.5 as a whole, it is with certainty. This clearly reflects that the overall certainty is contained in the individual uncertain phenomena. Some other similar phenomena were found in natural, which have the analogical regular patterns. In this chapter, it provides the stories that the overall certainty is presented though the relevant individual event is uncertain or incident.

8.2 Statistical Laws

1. **Galton knocked boards**

As shown in the Fig. 8.1, each black dot represents a nail nailed to the board. They are all at the same distance from each other, and the horizontal position of each one of the upper layers is just in the middle of the two of the lower layers. Put a small

M. Zheng and J. Yu, *Systems Theory for Engineering Practice*,
https://doi.org/10.1007/978-981-97-9342-6_8

round glass ball with a diameter slightly smaller than the distance between two nails from the entrance. When a small ball falls down, it rolls down to the left or right with a probability of 1/2 after hitting the nail, so it hits the next nail. This continues until it rolls into a grid on the bottom plate. Put many small balls of the same size down from the entrance continuously. As long as the number of balls is quite large, they will pile up on the bottom plate to form an approximate normal density function graph, that is, the center is high and the two ends are low, which shows an ancient bell shape with left and right symmetry.

In this experiment, although the prediction of final position of each small round glass ball is impossible, i.e., uncertain event, the total distribution of all small round glass balls will be with an approximate normal density function graph centered around the central point of the board as a whole provided that a large number of balls are used.

2. **Light diffraction**

Diffraction occurs whenever a propagating wave meets an obstacle in its path to limit its propagation by blocking or absorbing part of the front [1]. After the obstacle, the trajectories of each small segment of the front no longer propagate in the original direction. Such phenomenon appears in propagation of every type of wave. They all have the similar mathematical descriptions though with different physical aspects from case to case. When the size of the obstacle is on the same order of magnitude as the wavelength, diffraction occurs.

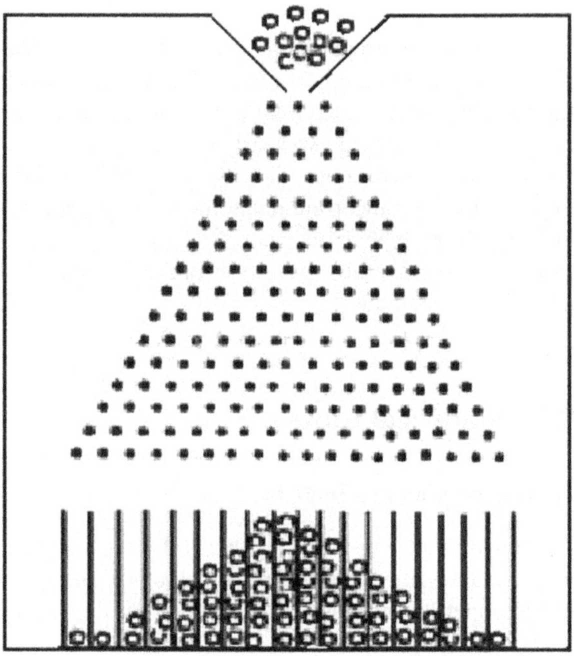

Fig. 8.1 Galton knocked boards

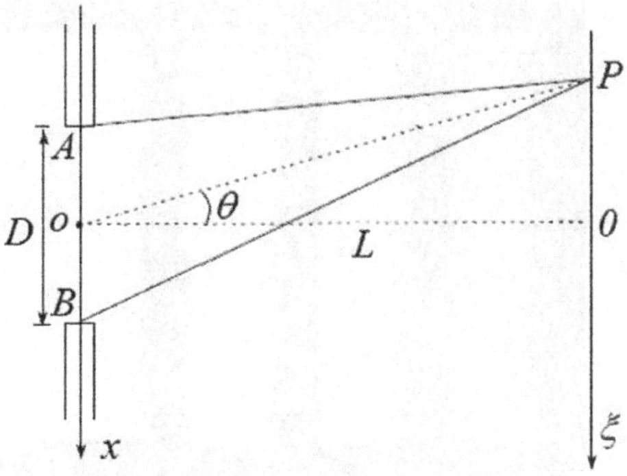

Fig. 8.2 Schematic of the Fraunhofer diffraction of a slit

As to the diffraction of light wave, it can be shown in the diffraction by the slit in Fig. 8.2. The propagation directions of the elements of the diffracted wave are in parallel.

Figure 8.3 shows the intensity distribution of light on the screen ahead of the slit with a distance L, in which ξ indicates the coordinate on the observation screen parallel to x direction and with origin at the point corresponding to $\theta = 0$. Here the size D of the slit is on the same order of magnitude as the wavelength λ. Figure 8.4 shows the diffraction image of light on the screen.

Surely, Fig. 8.4 displays a high central maximum, and sequences of alternated maxima and minima symmetrically on the two sides. Obviously, the intensity distribution of light after the slit shows the overall distribution of light wave (or a lot of photons) projected on the screen which interacted, it is deterministic as a whole though projection position of each "photon" is unknown.

3. **Electron diffraction**

Electron beam diffraction is an experimental technique, in which electron beam is projected onto a sample material to observe patterns against interference [2]. This is due to the dual nature of particles and waves of electrons. Electrons (particles) can also be seen as waves due to their characteristic of wave behavior. Therefore, electrons can be regarded as the common waves like sound waves or water waves, etc. Similar

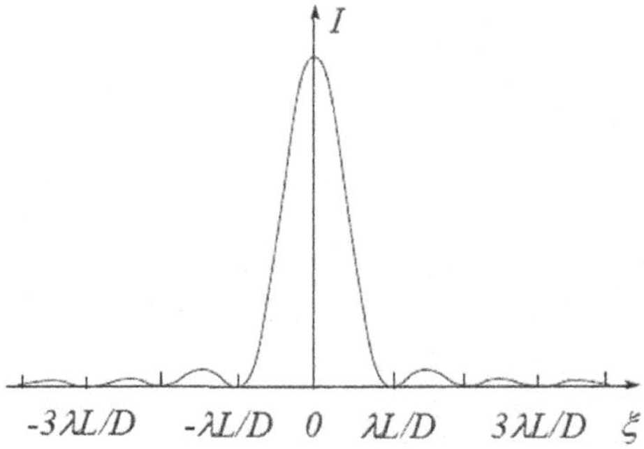

Fig. 8.3 Intensity curve in diffraction of a slit under Fraunhofer conditions

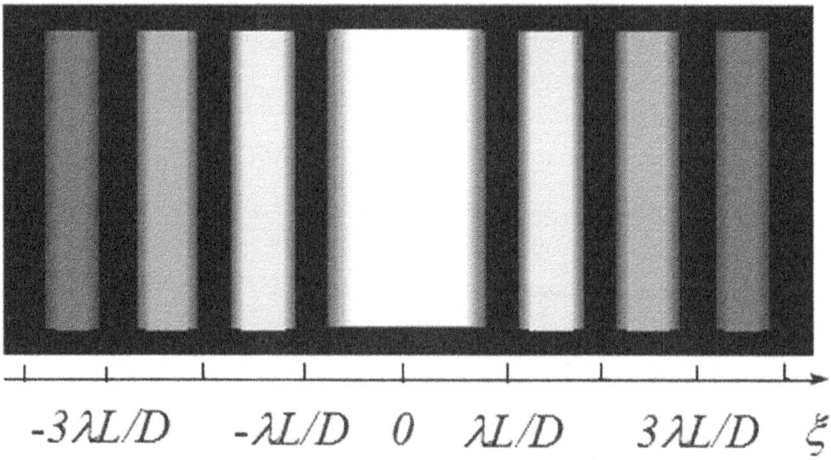

Fig. 8.4 Diffraction fringes of a slit under Fraunhofer conditions

techniques include X-ray diffraction and neutron diffraction, electron diffraction is usually used to study the crystal structure of solids in solid state physics and chemistry. The most typical experimental equipment that can produce electron diffraction (confined field electron diffraction or SAED) patterns is transmission electron microscope (TEM). There is also scanning electron microscope (SEM) equipped with detectors, which can acquire electron backscattered diffraction (EBSD) patterns. In both TEM and SEM, electrons are accelerated by electrostatic potential to obtain the required energy, which is set to have a specific wavelength, and then irradiated on the target sample.

Since the crystal has a periodic structure, it acts as a diffraction grating and scatters electrons in a predictable way. The observed diffraction is based on the pattern caused by lattice, so the Brave lattice can be determined from diffraction pattern. The crystal structure can be estimated by accurately measuring the diffraction intensity in principle, as in the case of X-ray diffraction. Figure 8.5 shows the pattern of electron beam diffraction by Cu_3Au alloy.

4. **Benford's law**

Benford's law [3], also known as Benford's rule, shows that the probability of the number with value of 1 as the first number in data is statistically about 30% of the total in a pile of data obtained from real life, the probability of the number with value of 2 as the first number in data is about 17.6%, ... Generally speaking, the larger the number, the lower the probability of the number is as the first digit. This rule was discovered by physicist Frank Benford in 1938. Benford's Law states that in the b-decimal system, the probability of numbers starting with the number n is expressed as $p = \log_b[(n+1)/n]$. For example, when $n = 1$, the corresponding probability is $p = \log_{10} 2 = 0.3010$ in 10-decimal system, that is, the probability of data with first number as 1 is 30.1%; when $n = 2$, the corresponding probability is

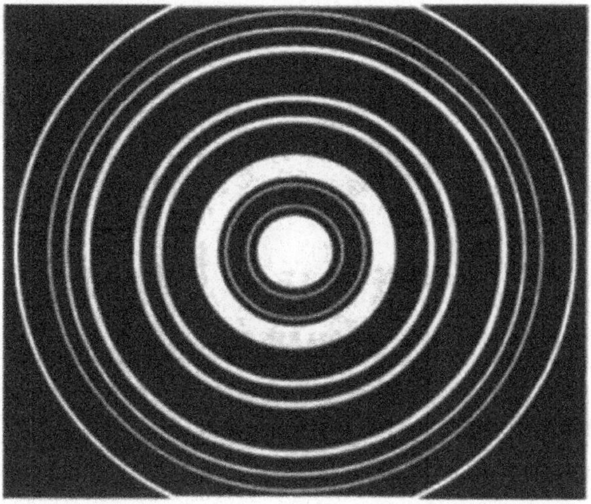

Fig. 8.5 Pattern of electron beam by Cu_3Au alloy

$p = \log_{10}[3/2] = 0.1761$ for the first digit of 2; ...; when $n = 9$, the corresponding probability is $p = \log_{10}[10/9] = 0.0458$ for the first digit of 9. Benford's law is not only applicable to single-digit numbers, but also to multi-digit numbers. It can be used to check whether all kinds of data are fraudulent. Besides, in 2009 Spanish mathematicians discovered a new pattern in prime numbers. Although prime numbers are generally considered to be randomly distributed, Spanish mathematicians have found that the first digit of each prime number in the prime number sequence has an obvious distribution law, which can be described by Benford's law of prime numbers. This new discovery not only provides a new insight into the properties of prime numbers, but also can be applied to fraud detection and stock market analysis.

In fact, counting is starting from 1 in sequence, 1, 2, 3, ... 9, if it ends here, all numbers seem to have the same chance, but after 9, from 10 to 19 it starts with 1, and the proportion of numbers headed by 1 increases greatly. Before it appears with the number starting with 9, it will inevitably go through a series of numbers starting with 2, 3, 4, ... 8. Therefore, the probability of numbers headed by 1 is generally much greater than that of numbers headed by 9, and in natural distribution, the probability of numbers headed by 1 to 9 decreases in turn. The natural distribution state is shown in Fig. 8.6.

Conditions for using Benford's Law

First, data must span large enough and must span several orders of magnitude to produce this result. Data samples should be as many as possible, at least more than 3000; simultaneously, the span of data samples should be large, for example, the height of people does not meet the "Benford's law" because most people are in the range of 1–2 m. Second, some manmade data does not meet this law, such as mobile phone number, ID number, invoice number and other data, which obviously does not meet this logarithmic distribution law. In other words, Benford's law is a law that is

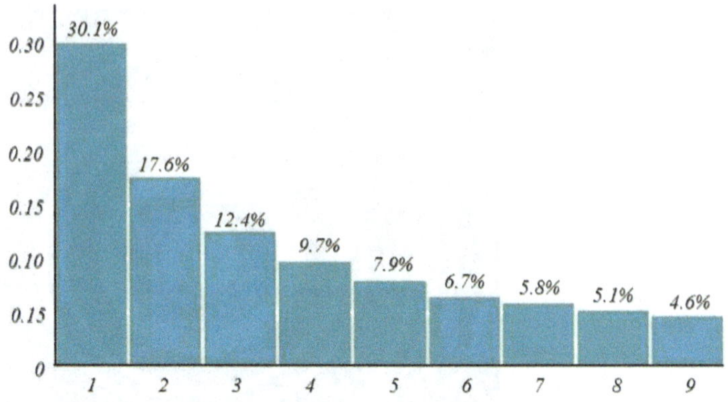

Fig. 8.6 Natural distribution state of number

revealed without any restrictions. The more artificial restrictions are placed on the generation of data, the less satisfied it is. Third, the data can't be artificially modified, and the artificially modified data are generally not satisfied with Benford's law.

5. **Gutenberg-Richter law**

Gutenberg-Richter Law works in seismology. It was put forward by seismologists Bino Gutenberg and Charles Francis Richter in 1956. The law reflects the relationship between magnitude and the number of earthquakes greater than or equal to a specific magnitude in a certain region [3]. The expression of this law is

$$\log_{10} N = a - bM\,, \text{or } N = 10^{(a-bM)}. \tag{8.1}$$

In Eq. (8.1), N represents the number of earthquakes with magnitude greater than or equal to M; a and b are constants. Figure 8.7 displays the diagram of N *versus* M.

Fig. 8.7 Diagram of N versus M

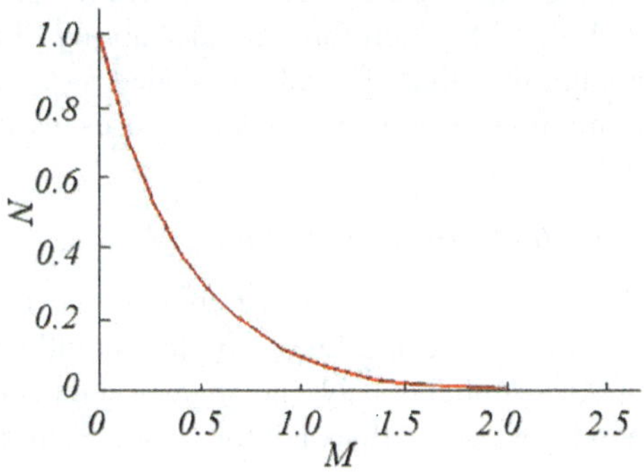

8.3 Schrodinger's Cat and Its Enlightenment

"Schrodinger's cat" is a famous thought experiment about the superposition of cat life and death states, which was put forward by Austrian physicist Schrodinger in 1935 [4, 5]. It is a deduction to extend the quantum behavior in the micro field to the macro world, which recommends people to understand a phenomenon of quantum behavior: observation. Microscopic matter has different forms of existence, namely particles and waves. Usually, microscopic matter exists in the superposition chaotic states of waves; once observed, they immediately choose to become particles. The experiment is like this: a poor cat is kept in a box, and the release of toxic gas controlled by radioactive substances is also placed in the box. After that, the cat will be poisoned because the radioactive material will decay with a 50% probability in a certain time, and the cat will live with a 50% probability that the radioactive material will not decay.

According to classical physics, one of these two results is bound to happen in the box, and the external observer can only know the results inside by opening the box [4, 5]. In the quantum world, when the box is closed, the whole system keeps the superposition state of uncertain waves, that is, the superposition state of the cat's alive and dead states. Whether the cat is dead or alive can only be determined after the box is opened and the matter is expressed in the form of particles when external observers observe it. The purpose of this experiment is to demonstrate quantum mechanics' extraordinary knowledge and understanding of the world of microscopic particles, but this makes the principle of microscopic uncertainty become the principle of macroscopic uncertainty, the objective law is not changed by human's willing, and the state that cat is alive and dead seems to violate logical thinking.

Schrodinger's cat itself is a hypothetical thought experiment. With the development of technology, people have realized the Schrodinger cat state in photons, atoms and molecules, and even have begun to try to prepare the Schrodinger's cat state with viruses. On the other hand, it is found that Schrodinger's cat state (quantum superposition state) itself exists in the process of life and is indispensable for biological survival.

Even though the experiment of Schrodinger's cat tells us that the macro world we live in is a deterministic world, and every event has a definite result. But in the quantum world, the state of things exists with probability. Besides, this experiment tells us that we must accept that some things exist in probability, not completely certain. We need to learn to face uncertainty, accept the possibility of some things, and be prepared for what may happen. In daily life, people always try to find certain things to provide us with security and stability. However, when we think about it carefully, we find that there are many uncertain factors in this world, and we cannot guarantee that any decision is clear and reliable. Therefore, we need to improve our cognitive ability and accept this uncertainty. The state of Schrodinger's cat has aroused people's new understanding of the definition of things. Just like this cat in a box, its living state is not only death or alive, but also includes both, each of which exists with a certain probability. The definitions of "living" and "death" become vague and unclear in this state, which enlightens people to look at the adjustment point of view again.

Quantum entanglement is a very unique and wonderful phenomenon in quantum mechanics, which refers to the phenomenon that when several particles interact with each other, the properties of each particle have been integrated into a whole, so it is impossible to describe the properties of each particle (sub-system) separately, but only the properties of the whole system. It describes the state of interdependence and connection between two or more quantum sub-systems. In this state, no matter how far they are, when one sub-system changes, the other sub-system will make response and corresponding changes immediately. This connection seems to go beyond the way that classical physics can explain. This phenomenon has caused a lot of discussions about whether the speed of information transmission will exceed the speed of light, and questioned our understanding of the nature.

The main subtle problems and doubts involved in quantum entanglement are as follows. First of all, what causes quantum entanglement? At present, scientists believe that when two or more systems interact, their states will become interrelated. For example, when two particles collide, the interaction between them will cause their states to get entangled. However, the specific mechanism and reasons are not completely clear, which is still a problem to be solved. Secondly, how does quantum entanglement realize information transmission? Although quantum entanglement can make the connection between the two systems instantly, it can't realize superluminal information transmission. Finally, does quantum entanglement exist only in the microscopic world or can it be extended to the macroscopic world? In a word, quantum entanglement is an important concept of quantum mechanics, which is used to describe the state of interdependence and connection between two or more quantum systems. Although we have some in-depth understanding of quantum entanglement, there are still some subtle problems and doubts which need to be further studied and answered. With the development of science, we believe that the understanding of quantum entanglement will be deeper and deeper, which will bring us more enlightenment about the nature.

8.4 Parallels Between Modern Physics and Eastern Mysticism

F. Capra has been devoted to find the parallels between modern physics and Eastern mysticism, as well as environmental activism and sustainability of society, for several decades [6, 7]. In his words, the "mysticism", at least in the west, has been associated traditionally, quite erroneously, with things vague, mysterious, and highly unscientific [6, 7]. However, this situation is now changed fortunately. A large number of people are interested in Eastern thought, the mysticism is taken even within the scientific community seriously.

Modern physics produced a significant effect on our society, which solidified the fundamental of natural science and technical science, and changed the conditions of life on our earth basically, in both beneficial and detrimental ways. Today, almost all industry makes use of the achievements of atomic physics, which influences the political structure of the world through their application, this means that the influence of modern physics goes beyond sole technology [6, 7]. Such phenomenon attracted discussions by both physicists and philosophers widely. Capra found that the concepts of modern physics often show surprising parallels to the ideas expressed in Eastern mysticism [6, 7].

The Tao of Physics changed his life dramatically. His professional career has been broadened from a physicist of particle physics to a systems theorist, philosopher of science, etc. He traveled extensively around the world to discuss philosophical and social implications of contemporary science with professional and outsider listeners for many years. In Capra's view [6, 7], the universe can no longer be seen as a machine that consists of elementary building blocks. While the material world is a network of inseparable patterns of relationship ultimately; the planet as a whole is a living, self-regulating system. The human body is not only the brain but also the immune system, the tissues bodily, and even each cell as a living, cognitive system instead of a machine and of the mind as a separate entity. Evolution is a cooperative dance with creativity and the constant emergence of novelty as the driving forces instead of a competitive struggle for existence. This is a new view on complexity, networks, and patterns of organization.

It is realized that the constituents of matter and the basic phenomena are all interconnected, which could be understood as integral parts of a unified whole instead of isolated entities. This is one of the strongest parallels to Eastern mysticism. The notion of a basic "quantum interconnectedness" got renewed attention these decades due to the realization of the universe being interconnected in much subtler ways.

So, *perhaps the set of L–V (Lotka–Volerra)-like primary system of dynamic equations formally proposed in Chap.* 1 *to characterize the interaction among the "five essential elements" is helpful to derive appropriate solutions to the problems with complex interconnections* in a sense.

The most impressive developments in particle physics might be *S*-matrix theory and the bootstrap approach [6, 7], which tries to understand nature entirely through its self-consistency instead of adopting any fundamental entities. The bootstrap philosophy was considered to be the culmination of current scientific thinking, it was seen closest to Eastern thought [6, 7].

The *S*-matrix approach is holistic and thoroughly dynamic, which sees particles as interrelated energy patterns in an ongoing universal process—as correlations, or interconnections, between various parts of an inseparable cosmic web. In such a framework, the meaning of *quark structure* is to reflect the fact of the transfer of energy and the flow of information in this network of events proceeding along well-defined lines, which creates the two-ness associated with mesons and the three-ness associated with baryons. This is the dynamic equivalent to the statement of hadrons consisting of quarks. In *S*-matrix theory, there is only a flow of energy showing certain well-defined patterns without distinct entities and basic building blocks [6, 7].

In subatomic physics, local variables are represented by connections between two spatial-separated events through signals, and the limit transmission speed of signal is the speed of light. However, beyond these local connections, nonlocal connections have emerged recently. Although the nonlocal connections are instantaneous and cannot be accurately predicted in a mathematical way, they are seen as the very essence of quantum reality by some physicists. In quantum theory, an individual event does not always have a well-defined cause. It means that the exhibition of any part is determined by its nonlocal connections to the whole. So, one has to employ the wider concept of statistical causality to replace the narrow classical notion of cause and effect. The laws of atomic physics are statistical laws [6, 7], the probabilities of atomic events are determined by the whole system's dynamics. It means that the whole determines the behavior of the parts in quantum physics, which is completely reversed in classical physics.

In the classical physics and macroscopic world, nonlocal connections are relatively unimportant, which lead to the separate objects and the formulation of the laws to describe their behavior in terms of certainties meaningful. But *in smaller dimensions, the influence of nonlocal connections becomes significant, the certainties give way in probability manners, which result in more and more difficult to separate any part of the universe from the whole*. Furthermore, Capra put forward new paradigm with six criteria for science [6, 7].

The first criterion is the relationship between the part and the whole, which is with more symmetrical. Not only the properties of the parts certainly are indispensible to the understanding of the whole, but also the properties of the parts can be understood fully by knowing the dynamics of the whole at the same time.

The whole is a primary integral, if you know the dynamics of the whole, you could in principle derive the patterns and properties of interactions of the parts. This view of the relationship between the part and the whole was recognized in quantum theory as a new creation [6, 7]. It is even realized that parts could no longer be well defined. The different properties can be seen in specific experimental context.

At the atomic level, the mechanical universe that composed of fundamental building blocks does not appear, which is replaced by a network of relations. All things are seen as interdependent, inseparable, and as transient patterns of the same ultimate reality.

The 2nd criterion is the *thinking in terms of process* instead of thinking in terms of structure. The process is primary, the every apparent structure observed is a manifestation of an underlying process. Process thinking is also a main characteristic of Eastern mystical traditions.

The 3rd criterion is the replacement of image of the universe as a machine by that of an interconnected, dynamic whole. The parts of the interconnected—dynamic whole are essentially interdependent.

The 4th criterion is about the building of knowledge. The foundations of scientific knowledge have not always remained solid, but developed with awareness.

The 5th criterion is about the approximate descriptions of science. *All scientific concepts and theories are limited and approximate.* Science can never provide any complete and definitive understanding. Scientists can deal with limited and approximate descriptions of reality.

The 6th criterion is about the advocacy on observation. It is advocated the *shift from an attitude of domination and control of nature*, including human beings, *to one of cooperation and nonviolence.*

8.5 Goodhart's Law

Goodhart's law, named after Charles Goodhart, is a very famous theorem, which states that *when a policy becomes a goal, it will no longer be a good policy* [3]. As a former adviser of the Bank of England, it is proposed that when the government tries to manage the special signs of these financial assets, they are no longer credible economic indicators. He first proposed it in an article in 1975, and later became an important basis for criticizing Britain's expansionary or tight monetary policy. This law can be interpreted in many ways, and the most common one is that once a social indicator or economic indicator becomes an established goal to guide macro-policy making, it will lose its original information value. In fact, *if policy makers will sacrifice other aspects to strengthen this indicator, this indicator no longer has the role of indicating the overall situation.* To put it another way simply, once *an indicator becomes the basis for policy making, it will no longer be effective.* Policymakers will strengthen this indicator at the expense of other aspects, so that this indicator no longer has the function of indicating the overall situation.

There was also a joke, a woman keeps a hen and feeds her barley every day. The hen gives her an egg every day in response. One day, the woman thought, wouldn't it be better to give me two eggs if I doubled the hen's food intake? So she doubled the barley she fed the hen. However, hens are getting fatter and fatter. Finally, she stopped laying eggs. As you sow, you reap, which is a simple linear thinking. Because of the *complex interaction in the system, its function is entangled and nonlinear*, so *it can not be solved simply by linear method.*

In China, there is an idiom "*things will develop in the opposite direction when they become extreme*", this is nature and also a fundamental principle of Eastern philosophy.

The relevant philosophy and dialectics, such as laws, idiom and histories, could not only enrich knowledge but also activate wisdom and its transformation to productivity.

8.6 Ranking of Fuzzy Numbers

For general precise (exact, non-fuzzy) numbers, it is very easy to compare their sizes, and it can be determined according to their position order of two numbers on the real number axis. However, for fuzzy numbers, it is not so easy to judge two fuzzy numbers, [8, 9]. Since 1970s, people began to study how to determine the ranking of fuzzy numbers, and put forward more than 20 methods. Nevertheless, none of them has been widely accepted [8, 9]. Recently, a new ranking method of fuzzy numbers was proposed on basis of probabilistic multi-objective optimization in viewpoint of systems theory [10]. In this section it is outlined with an example of optimal option from four candidate schemes with one performance.

Analogical to the method of "center of gravity" in mechanics, the mean value (M) and the standard deviation (σ) of the performance in respect of membership of fuzzy number $\tilde{A}$ can be evaluated by using Eqs. (8.2) and (8.3), respectively [8, 9],

$$M = \int\limits_{x \in R} x\mu(x)dx / \int\limits_{x \in R} \mu(x)dx, \tag{8.2}$$

$$\sigma = \left[\int\limits_{x \in R} (x - M)^2 \mu(x)dx / \int\limits_{x \in R} \mu(x)dx] \right]^{0.5}. \tag{8.3}$$

In Eqs. (8.2) and (8.3), $\mu(x)$ indicates the membership of fuzzy numbers, and $\int_{x \in R} \mu(x)dx \neq 0$.

For triangle type fuzzy number (l, m, u), as is shown in Fig. 8.8 by Form 1, it derives [9],

$$M = (l + m + u)/3, \sigma = \left[(l^2 + m^2 + u^2 - lm - lu - mu)/18 \right]^{0.5}. \tag{8.4}$$

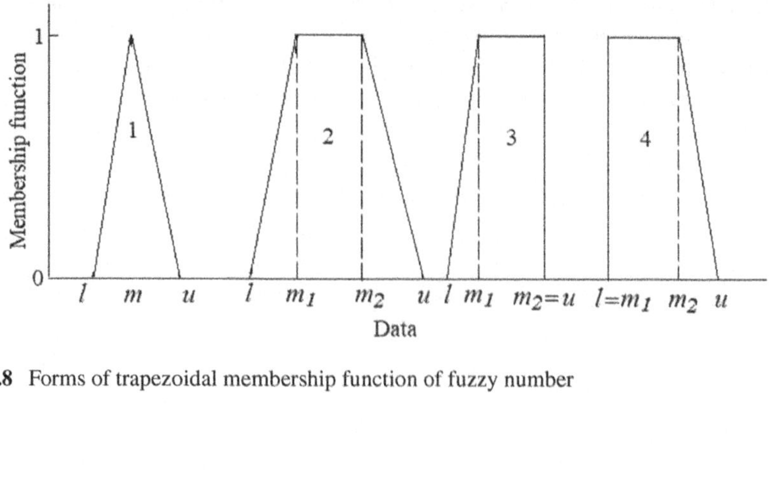

Fig. 8.8 Forms of trapezoidal membership function of fuzzy number

For trapezoidal type fuzzy number (l, m_1, m_2, u), as is shown in Fig. 8.8 by Form 2, it gives [9],

$$\begin{aligned} M &= \left[\left(u^2 + m_2^2 + um_2 - m_1^2 - l^2 - lm_1\right)\right]/[3(u + m_2 - l - m_1)], \\ \sigma &= \{\left[\left(m_1^4/4 + l^4/12 - lm_1^3/3\right)/(m_1 - l) + \left(m_2^3 - m_1^3\right)/3 \right. \\ &\quad \left. + \left(u^4/12 + m_2^4/4 - um_2^3\right)/(u - m_2)\right] \\ &\quad /[(u - m_2)]/[(u + m_2 - l - m_1)/2] - M^2\}^{0.5}. \end{aligned} \tag{8.5}$$

Through above proceldure, the fuzzy numbers thus transfer into *defuzzied* ones by means of the mean value (M) and the standard deviation (σ). Thus fuzzy ranking of fuzzy numbers problem now becomes a dual-objective optimization problem in cases of "the bigger the better" in respect of the *defuzzied* mean value (M) and the standard deviation (σ) of the performance per scheme, thus the general probabilistic multi-objective optimization can be used [10].

Now, let's restudy Li's comprehensive data of scheme selection by optimal comparison as an example [9], the comprehensive data of evaluation of the four alternative schemes of flights with triangle type fuzzy number are cited in Table 8.1.

The values of the mean values and standard deviation of fuzzy numbers $\widetilde{A}$, $\widetilde{B}$, $\widetilde{C}$ and $\widetilde{D}$ are cited in Table 8.2 [9].

Furthermore, according the probabilistic multi-objective optimization, the parts of partial preferable probabilities and entire/total preferable probabilities could be evaluated, which are shown in Table 8.3 [10]. In the evaluations, the mean value of the comprehensive data is beneficial type of performance index according to its role, the standard deviation is unbeneficial type of performance index in general [10].

Table 8.3 indicates that the rank of these four types of flights is $A > C > D > B$ according to the value of their total preferable probabilities of the four alternative schemes of flights in viewpoint of systems theory.

However, in Li's evaluation [9], the ranking criteria are shown in Table 8.4, which doesn't take the role of the mean value M and the standard deviation σ of fuzzy

Table 8.1 Comprehensive data of evaluation of the four alternative schemes [9]

Alternative scheme	A	B	C	D
Fuzzy value of comprehensive evaluation	$\widetilde{A}$ = (0.2, 0.3, 0.5)	$\widetilde{B}$ = (0.1, 0.35, 0.6)	$\widetilde{C}$ = (0.25, 0.4, 0.7)	D = (0.15, 0.2, 0.4)

Table 8.2 Mean values and standard deviations of the four alternative schemes [9]

Alternative scheme	A	B	C	D
Mean value	0.333	0.35	0.45	0.25
Standard deviation	0.062	0.102	0.094	0.054

Table 8.3 Partial and total preferable probabilities of the four flights

Alternative scheme	A	B	C	D
Part of partial preferable probability of M, P_M	0.2408	0.2531	0.3254	0.1808
Part of partial preferable probability of σ, P_σ	0.3013	0.1731	0.1987	0.3269
Entire/total preferable probability, $P_t \times 10^2$	7.2543	4.3801	6.4659	5.9097
Rank	1	4	2	3

Table 8.4 Ranking criteria in respect of the mean value M and the standard deviation σ of fuzzy numbers $\widetilde{A}$ and $\widetilde{B}$ for a performance [9]

Relationship between $M(\widetilde{A})$ and $M(\widetilde{B})$	$M(\widetilde{A}) > M(\widetilde{B})$	$M(\widetilde{A}) = M(\widetilde{B})$		
Relationship between standard deviations	No matter	$\sigma(\widetilde{A}) < \sigma(\widetilde{B})$	$\sigma(\widetilde{A}) > \sigma(\widetilde{B})$	$\sigma(\widetilde{A}) = \sigma(\widetilde{B})$
Ranking	$\widetilde{A} > \widetilde{B}$	$\widetilde{A} > \widetilde{B}$	$\widetilde{A} < \widetilde{B}$	$\widetilde{A} = \widetilde{B}$

number $\widetilde{A}$ as a whole in the system at the same time. Subsequently, Li gave different evaluation results for the data of Tables 8.1 and 8.2 with criteria of Table 8.4, which is $C > B > A > D$ [9], it differs from our result distinctly.

Obviously, in the viewpoint of systems theory and algorithm of probability theory, in term of overall/whole optimization, the ranking evaluation of fuzzy numbers $\widetilde{A}$ and $\widetilde{B}$ could be conducted by simultaneously considering both the mean value M and the standard deviation σ of a fuzzy number $\widetilde{A}$ by means of preferable probability as the overall assessment, and the ranking could be determined according to the total preferable probability [10].

8.7 Summary

Although individual uncertain event is an unpredictable incident, the overall certainty of the system as a whole exists undoubtedly; the "unity of opposites" occurs inside a system deterministically. Moreover, "things will develop in the opposite direction when they become extreme" unavoidably due to the complex nonlinear in the system usually. The wholeness of things leads to the objective laws of the real world, and obliges us to recognize the world with a global view.

References

1. A. Bettini, *A Course in Classical Physics 4—Waves and Light*, (Springer, Cham, Switzerland, 2017)

2. S.X. Zhou, H. Chen, *A Course in Quantum Mechanics*, 2nd edn. (Higher Education Press, Beijing, 2009)
3. Kit Yates, *How to Expected: The Science of Making Predictions and the Art of Knowing When not to, Quercus Editions Limited, through The Grayhawk Agency Ltd.* (CITIC Press Corporation, Beijing, 2024)
4. J.Y. Zeng, *Quantum Mechanics*, vol. 2, 5th edn. (Science Press, Beijing, 2014)
5. J.Y. Zeng, *A Course of Quantum Mechanics*, 3rd edn. (Science Press, Beijing, 2014)
6. F. Capra, *The Tao of Physics, an Exploration of the Parallels Between Modern Physics and Eastern Mysticism*, 5th edn. (Shambhala, Boston, 2013)
7. F. Capra, *The Turning Point, Science, Society and the Rising Culture.* (Simon and Schuster, New York, 1982)
8. S.J. Chen, C.-L. Hwang, *Fuzzy Multiple Attribute Decision Making, Methods and Applications.* (Springer, Berlin, Heidelberg, 1992). ISBN-13: 978-3-540-54998-7, e-ISBN-13: 978-3-642-46768-4. https://doi.org/10.1007/978-3-642-46768-4
9. D.F. Li, *Fuzzy Multiobjective Many-person Decision Makings and Games* (Defense Industrial Press, Beijing, 2003). ISBN 7-118-02974-2
10. M. Zheng, J. Yu, *Robust Design and Assessment of Product and Production by Means of Probabilistic Multi-objective Optimization.* (Springer, Singapore, 2024)

Chapter 9
Typical Systemic Approaches in Broad Fields

Abstract This chapter gives typical systemic approaches in broad fields, such as, Lotka-Volterra system of ecology, "five essential elements" system of ancient China, method of electric images in electrostatics, and nonlocal Euler–Bernoulli cantilever in continuum mechanics, which reflect the universality of systemic phenomena and the relevant analysis.

Keywords Systems theory · Ecology · Five essential elements · Method of electric images · Nonlocal Euler–Bernoulli cantilever · Synergetics · Wholeness

9.1 Introduction

Many performances and aspects behind natural phenomena are correlated synergetically in viewpoint of systems theory. In ecology, American scholar Lotka and Italian scientist Volterra proposed the so called *Lotka-Volterra competitive models* in 1925 and 1926 individually, as well as predator–prey models, which can be seen as the basis of ecology. "Five essential elements" system is the earliest systemic viewpoint of ancient Chinese to see the nature, which was developed as a philosophic viewpoint to recognize and analyze the things in the worlds as a whole synergetically, especially in the traditional Chinese medical science. Method of electric images in electrostatics is an efficient approach to describe the synergetic action between electric charge and the surrounding conductors in the system, which ingeniously adjusts the distribution of induced charge in the conductors to satisfy the boundary conditions coordinately. The intrinsic characteristic of nonlocal theory is to consider the collective actions of material elements within the body, it is actually a "systemic view" in some senses. The nonlocal Euler–Bernoulli beam theory has received much attention in micro- and nano-devices recently due to the rapid development of nanotechnology.

In this chapter, these typical examples are presented to reflect the universality of systemic phenomena and the relevant analysis.

M. Zheng and J. Yu, *Systems Theory for Engineering Practice*,
https://doi.org/10.1007/978-981-97-9342-6_9

9.2 Lotka-Volterra System of Ecology

Predator–prey system is one of the basic components of ecosystem, and their interactions have a significant impact on maintaining ecological stability. Many biological populations play the roles of predator and prey at the same time, that is, together with other biological populations, they form a huge and complex food chain. There have been a lot of research results on food chain models with three or more species. Experimental and theoretical studies show that even the food chain system with extremely simple structure will show rich dynamic characteristics. All these properties show that there are very complex interactions and influences between species, and changes in the conditions of one species may cause changes in the number of other populations, which will affect the whole body.

Consider an ecosystem composes of two kinds of animals, one of which is preyed by the other. Let n_1 represent the number of prey animals, and n_2 express the number of predator animals. For a certain moment t, there is following Lotka-Volterra formulae [1, 2],

$$\begin{cases} \dot{n}_1 = dn_1/dt = r_1 n_1[1 - a(n_1 + bn_2)]; \\ \dot{n}_2 = dn_2/dt = r_2 n_2[1 - c(n_2 + dn_1)]. \end{cases} \tag{9.1}$$

In Eq. (9.1), the parameters r_1, r_2, a, b, c and d express the mutual influence coefficients, which determine the competitive result of species, which are with positive values.

Under equilibrium condition of the two species, i.e., $\dot{n}_1 = \dot{n}_2 = 0$, it derives,

$$\begin{cases} n_1[1 - a(n_1 + bn_2)] = 0; \\ n_2[1 - c(n_2 + dn_1)] = 0. \end{cases} \tag{9.2}$$

One group of the solutions of Eq. (9.2) is that $n_1' = n_2' = 0$, which is a plain solution without any practical physical meaning. While the other group of the solutions of Eq. (9.2) is that $n_1^* = \frac{1}{a} - \left(\frac{1}{c} - \frac{d}{a}\right) \cdot \frac{b}{1-bd}$, and $n_2^* = \left(\frac{1}{c} - \frac{d}{a}\right) \cdot \frac{1}{1-bd}$, , which is a specific solution (status). Around the special status $(n_1{}^*, n_2{}^*)$, if there appear small variations (fluctuations) ξ and ψ for n_1 and n_2, respectively, it derives,

$$\begin{cases} \dot{\xi} = -ar_1 n_1^* \cdot (\xi + b\psi); \\ \dot{\psi} = -cr_2 n_2^* \cdot (\psi + d\xi). \end{cases} \tag{9.3}$$

The solution of Eq. (9.3) indicates that both functions ξ and ψ decay with time parameter t by following expressions,

$$\begin{cases} \xi = Ae^{-\lambda t}; \\ \psi = Be^{-\lambda t}. \end{cases} \tag{9.4}$$

The parameter λ could take a definitely positive value given by Eq. (9.5),

$$\lambda = \frac{ar_1n_1^* + cr_2n_2^* + [(ar_1n_1^* - cr_2n_2^*)^2 + 4ar_1n_1^*cr_2n_2^*bd]^{0.5}}{2}. \tag{9.5}$$

The exponential attenuation performances of fluctuations ξ and ψ with time parameter t indicates that the status (n_1*, n_2*) is stable, so it is a stable solution of the system. Simultaneously the ratio of parameters A to B takes following value,

$$R \equiv A/B = ar_1n_1^*b/(\lambda - ar_1n_1^*). \tag{9.6}$$

Above analysis implies that as long as the numbers of species n_1 and n_2 satisfy the specific correlations determined by parameters r_1, r_2, a, b, c and d the system will get stable. The internal synergetics or coordination of the system is the key factor to get a stable state.

9.3 "Five Essential Elements" System of Ancient China

As was stated in Chap. 1, the "five essential elements" (Wuxing) system is the earliest systemic viewpoint to see the nature in ancient China, which was a philosophic viewpoint to analyze and recognize the things of the worlds as a whole, there are complex mutual interactions among the five essential elements, i.e., "generation" and "restriction".

Furthermore, according to ecosystem model [1, 2], a set of L—V (Lotka–Volerra)—like primary system of dynamic equations was formally proposed to reflect the interactions among the "five essential elements" [3], i.e.,

$$\begin{cases} \dot{u}_1 = u_1(\varepsilon_1 + \eta_{15}u_5 - \gamma_{14}u_4 - \alpha_1u_1), \\ \dot{u}_2 = u_2(\varepsilon_2 + \eta_{21}u_1 - \gamma_{25}u_5 - \alpha_2u_2), \\ \dot{u}_3 = u_3(\varepsilon_3 + \eta_{32}u_2 - \gamma_{31}u_1 - \alpha_3u_3), \\ \dot{u}_4 = u_4(\varepsilon_4 + \eta_{43}u_3 - \gamma_{42}u_2 - \alpha_4u_4), \\ \dot{u}_5 = u_5(\varepsilon_5 + \eta_{54}u_4 - \gamma_{53}u_3 - \alpha_5u_5). \end{cases} \tag{9.7}$$

In Eq. (9.7), u_i (i = 1, 2, ..., 5) represents the representative quantity of i-th essential element; ε_i (i = 1, 2, ..., 5) indicates the self incremental factor of i-th essential element; η_{ij} ($i = 1, 2, \ldots, 5; j = [(i+4)$ mod 5] expresses the representative generation factor of j-th essential element to i-th essential element, in which $[(i+4)$ mod 5] expresses the integer remainder of number $(i+4)$ divided by 5; γ_{ij} ($i = 1, 2,$, $5; j = [(i+3)$ mod 5] indicates the representative restraint factor of j-th essential element to i-th essential element; α_i (i = 1, 2, ..., 5) indicates the representative repelling factor of i-th essential element in itself mutually. Till now we have not known the exactly quantitative values of above representative factors.

Specially, if the system is at its stable status, its dynamic equations for the "five essential elements" satisfies, $\dot{u}_1 = \dot{u}_2 = \dot{u}_3 = \dot{u}_4 = \dot{u}_5 = 0$. In this specific condition, following conditions are needed to meet,

$$\begin{cases} \varepsilon_1 + \eta_{15}u_5 - \gamma_{14}u_4 - \alpha_1 u_1 = 0, \\ \varepsilon_2 + \eta_{21}u_1 - \gamma_{25}u_5 - \alpha_2 u_2 = 0, \\ \varepsilon_3 + \eta_{32}u_2 - \gamma_{31}u_1 - \alpha_3 u_3 = 0, \\ \varepsilon_4 + \eta_{43}u_3 - \gamma_{42}u_2 - \alpha_4 u_4 = 0, \\ \varepsilon_5 + \eta_{54}u_4 - \gamma_{53}u_3 - \alpha_5 u_5 = 0. \end{cases} \tag{9.8}$$

The solutions of Eq. (9.8) reveal the synergetic and coordinate correlations among the "five essential elements" in stable condition in principle.

Now, let's study a very special case, i.e., $\varepsilon_1 = \varepsilon_2 = \varepsilon_3 = \varepsilon_4 = \varepsilon_5 \equiv \varepsilon_0$, η_{ij} $(i = 1, 2, \ldots, 5; j = [(i + 4) \bmod 5] = \gamma_{ij}$ $(i = 1, 2, \ldots, 5; j = [(i + 3) \bmod 5] = \alpha_i$ $(i = 1, 2, \ldots, 5) = 1$. In this case the particular solution is, $u_1{}^* = u_2{}^* = u_3{}^* = u_4{}^* = u_5{}^* = \varepsilon_0$, which is a particular state. Around the particular state $(u_1{}^*, u_2{}^*, u_3{}^*, u_4{}^*, u_5{}^*)$, if there appear small variations (fluctuations) ξ, ψ, μ, ν, and τ for u_1, u_2, u_3, u_4 and u_5, respectively, it derives following formulae under first approximation,

$$\begin{cases} \dot{\xi} = \varepsilon_0(\tau - \nu - \xi), \\ \dot{\psi} = \varepsilon_0(\xi - \tau - \psi), \\ \dot{\mu} = \varepsilon_0(\psi - \xi - \mu), \\ \dot{\nu} = \varepsilon_0(\mu - \psi - \nu), \\ \dot{\tau} = \varepsilon_0(\nu - \mu - \tau). \end{cases} \tag{9.9}$$

The solution of Eq. (9.9) indicates that the functions ξ, ψ, μ, ν, and τ decay with time parameter t by following expressions,

$$\begin{cases} \xi = Ae^{-\lambda t}; \\ \psi = Be^{-\lambda t}; \\ \mu = Ce^{-\lambda t}; \\ \nu = De^{-\lambda t}; \\ \tau = Ee^{-\lambda t}. \end{cases} \tag{9.10}$$

The parameter λ could take a definitely positive value given by Eq. (9.11),

$$\lambda = \varepsilon_0. \tag{9.11}$$

The exponential attenuation performances of fluctuations ξ, ψ, μ, ν, and τ with time parameter t indicates that the state $(u_1{}^*, u_2{}^*, u_3{}^*, u_4{}^*, u_5{}^*)$ is stable, so it is a stable solution of the system. Simultaneously the ratio of parameters *A, B, C, D and*

E takes following value,

$$A : B = B : C = C : D = D : E = E : A = 1. \tag{9.12}$$

This analysis indicates that the internal coordination or synegetics of the system is the key factor to get a stable state. As long as the "five essential elements" satisfy the specific correlations determined by parameters ε_i ($i = 1, 2, \ldots, 5$), η_{ij} ($i = 1, 2, \ldots, 5; j = [(i + 4) \bmod 5]$, γ_{ij} ($i = 1, 2, \ldots, 5; j = [(i + 3) \bmod 5]$ and α_i ($i = 1, 2, \ldots, 5$) the system could get stable.

An extremely special case is that ε_i ($i = 1, 2, \ldots, 5$) $= \varepsilon_0 = 0$, it derives a group of plain solution without any practical physical meaning, i.e., $u_1' = u_2' = u_3' = u_4' = u_5' = 0$.

Other solutions are valuable to dig.

9.4 Method of Electric Images in Electrostatics

Method of electric images in electrostatics is a very special approach that was used widely to solve electrostatic problems systemically [4–6].

The general procedure to ensure the boundary conditions met by the potential V of the electrostatic field is to use following criteria [4]:

(a) The potential is finite through the field, except for some locations where the point charges (sources) are placed;
(b) The potential is continuous in the whole field, which includes the surface of a conducting or dielectric body, except for double layers;
(c) On the surface of a conductor it needs to make sure either $V = V|_S =$ constant, or $\sigma = -\varepsilon(\partial V/\partial n)|_S$;
(d) At the interface between two dielectric media, if $\sigma = 0$, it has, $\varepsilon_1(\partial V/\partial n)|_1 = \varepsilon_2(\partial V/\partial n)|_2$;
(e) The point charges (sources) are distributed at a finite distance, so the potential is zero at positions of infinity distance.

This method could be used to determine some essential electric quantities (charge density, potential, force of interaction, etc.) for a few (one or more) point charges, in conditions of certain separation surfaces, e.g. maintained at a constant potential, or conductors connected to the ground.

The fundamental of method of images is the fact that a small number of point charges, and located in the geometry of the problem, can be chosen to simulate the boundary conditions required by the problem conveniently. These charges are called "image charges".

This method makes it possible to determine the charge density σ on the surface of a simply-shaped conductor (plane, sphere, etc.) coordinately, in condition of one or more point charges, which omits the real solving of Poisson's equation.

The core essence of such method is the uniqueness of the solution of Poisson's equation for given boundary conditions. A complicated configuration is modeled by a simple one provided that it has identical boundary conditions. As a result, the solution of the complicated problem is transferred into a solving of simple one with identical boundary conditions synergetically.

The distinct aspects of using this method are as follows:

1. Domain division

   The boundary conditions automatically separate the whole space into two domains, i.e., a domain of interest, D, in which the potential is to be determined, and a complementary domain, D', which is concealed behind the bounds. The method could give the correct potential in domain D only though incorrect potential in D', the latter is not concerned;
2. No additional charge or any other modification in domain D

   In order to ensure the Poisson's equation in the domain D to keep the same as the original problem, there is no additional charge or any other modification in domain D;
3. The additions of charges in complementary domain D' is to ensure the original boundary condition

   In the complementary domain D', one could add proper charges at chosen locations, or change permittivity, etc., to establish the new configuration in D' such that to produce the same boundary condition exactly at the interface between D and D' systemically.

Above requirements guarantee to keep the Poisson's equation with its boundary conditions as the original ones in D, which makes the problem in D easily to be solved by the additions of appropriate charges in the complementary domain D'.

A few examples to illuminate the procedure of the method of image charges will be given in the following sections.

Example 1. Plane Conductor As is shown in Fig. 9.1, an electric charge Q is placed at a position near an infinite plane conductor that is grounded, the distance from the electric charge Q to the conductor is a. The problem is to get the electric potential in the space [6].

Solution

Physical analysis indicates that there will have induced charges on the conducted plane. If the charge Q is positive, the induced charges will be negative. *The total electric field in space is generated by the charge Q and the induced charges on the plane conductor together.* On the other hand, *the distribution of induced charges gets their equilibrium status in conditions of actions of the total electric field.* The equilibrium condition is that the conductor plane is an equi-potential surface, i.e., the boundary condition of this problem is,

$$\varphi = const.\text{(on conductor's surface)}. \tag{9.13}$$

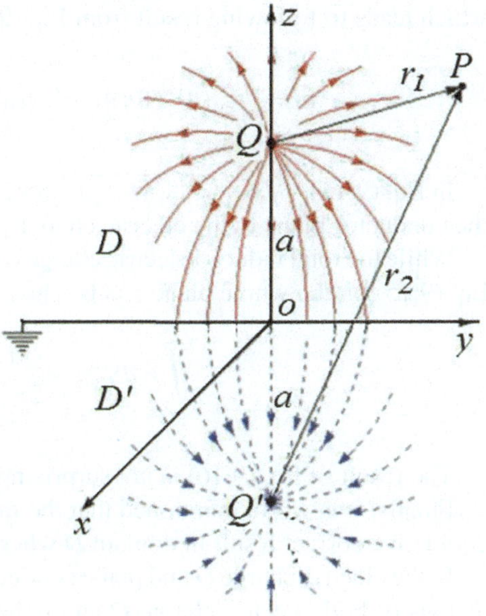

Fig. 9.1 Method of electric images for an electric charge near a plane grounded conductor

The direct solving of this problem is rather difficult.

An alternate approach is the method of electric images.

A hypothetical electric charge Q' is placed under the plane conductor (in domain D') with a distance a from the plane conductor. If the conductor plane is removed, the electric field in the space will be like that represented in Fig. 9.1 provided that $Q' = -Q$. The feature of such electric field is that the electric-power lines are perpendicular to the original conductor plane everywhere, therefore the equilibrium condition of the conductor plane being an equi-potential surface is satisfied. Thus, the real induced charges on the conductor plane can be replaced by the hypothetical electric charge Q', the latter is in the domain D', the electric charge Q' is the "image electric charge". Simultaneously, the electric charge in domain D is still the electric charge Q only.

Furthermore, the *electric filed in domain D is the superposition of both Q and* Q'. Obviously, these procedures make the problem in D easily to be solved by the additions of appropriate "image electric charges" in the complementary domain D'.

Thus, the electric field at point $P(x, y, z)$ in domain D can be written as,

$$\begin{aligned}\varphi(P(x,y,z)) &= \frac{1}{4\pi\varepsilon_0}\left(\frac{Q}{r_1} - \frac{Q}{r_2}\right)\\ &= \frac{1}{4\pi\varepsilon_0}\left(\frac{Q}{\sqrt{x^2+y^2+(z-a)^2}} - \frac{Q}{\sqrt{x^2+y^2+(z+a)^2}}\right).\end{aligned} \tag{9.14}$$

Simultaneously, the formula $\sigma = -\varepsilon_0(\partial V/\partial n)|_S = -\varepsilon_0(\partial V/\partial n)|_{z=0}$ can be used to assess the distribution of induced charges on the surface of the plane conductor, which leads to following result from Eq. (9.14),

$$\sigma = -\varepsilon_0(\partial V/\partial n)|_S = -\varepsilon_0(\partial V/\partial n)|_{z=0} = -\frac{Q}{2\pi}\frac{a}{r_0^3}. \tag{9.15}$$

In Eq. (9.15), $r_0 = \sqrt{x^2 + y^2 + a^2}$ indicates the distance from a point $(x, y, 0)$ on the conductor plane to the electric charge Q.

While the total induced electric charge on the plate could be derived by integrating Eq. (9.15) on the whole plane $z = 0$. Thus, it gets,

$$\int_S \sigma\, dS = -\frac{Qa}{2\pi}\int_S \frac{dS}{r_0^3} = -Q. \tag{9.16}$$

The result of Eq. (9.16) is not surprising.

Finally, it is worth mentioned that the method of electric images in electrostatics could give correct result in domain D where the original electric charge is placed.

In the electric charge Q and plane conductor system, the synergetic or cooperation actions of both electric charge Q and induced charge Q' is to ensure the boundary condition on the surface of the plane conductor, i.e., an equi-potential surface.

Example 2. Grounded Conducting Sphere An electric charge Q is placed near a grounded conducting sphere with radius R_0, see Fig. 9.2. The distance from the electric charge Q to the center of sphere is a ($a > R_0$). The problem is to find the electric potential in space [6].

Solution

Suppose that a hypothetical electric charge Q' is employed to substitute the induced electric charge on the grounded conducting sphere, the basic requirement is that the boundary condition of the electric potential on surface of the grounded conducting sphere, $\varphi = 0$ (constant), is met.

Considering the symmetry of the problem, the hypothetical electric charge Q' could be placed on the connected line of Q and O of the center of the sphere. The

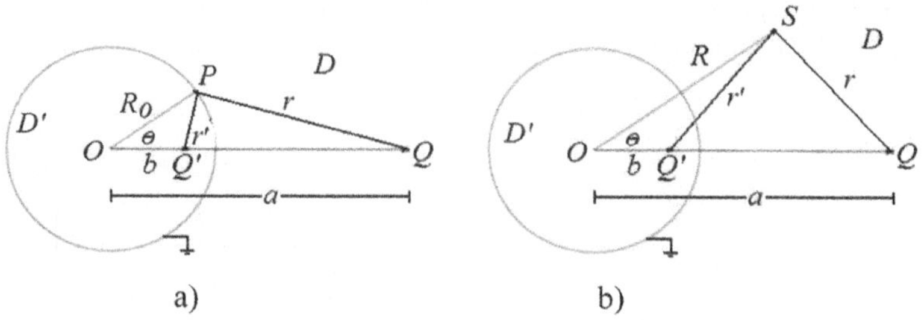

Fig. 9.2 The image problem of a charge near a grounded conducting sphere

distance from the hypothetical electric charge Q' to the center of sphere is b ($b < R_0$). In Fig. 9.2a, as to a point P on the grounded conducting sphere, the boundary condition of the electric potential on surface of the grounded conducting sphere is,

$$\varphi = \frac{1}{4\pi\varepsilon_0}\left(\frac{Q}{r} + \frac{Q'}{r'}\right) = 0. \tag{9.17}$$

In Eq. (9.17), r indicates the distance from Q to P, r' reflects the distance from Q' to P. The direct result of Eq. (9.17) is,

$$\frac{r'}{r} = -\frac{Q'}{Q} = const. \tag{9.18}$$

Detailed derivation gives [6],

$$b = \frac{R_0^2}{a}, \tag{9.19}$$

$$Q' = -\frac{R_0}{a}Q. \tag{9.20}$$

Equations (9.19) and (9.20) determine the value and position of the hypothetical electric charge Q'.

Thus for a point S in the domain D (outside the conducting sphere, Fig. 9.2b), the electric potential can be written as [5, 6],

$$\begin{aligned}\varphi &= \frac{1}{4\pi\varepsilon_0}\left(\frac{Q}{r} + \frac{Q'}{r'}\right) = \frac{Q}{4\pi\varepsilon_0}\left(\frac{1}{r} - \frac{R_0}{ar'}\right)\\ &= \frac{Q}{4\pi\varepsilon_0}\left(\frac{1}{\sqrt{R^2 + a^2 - 2Ra\cos\theta}} - \frac{R_0}{a\sqrt{R^2 + b^2 - 2Rb\cos\theta}}\right).\end{aligned} \tag{9.21}$$

In Eq. (9.21), r indicates the distance from Q to S, r' reflects the distance from Q' to S, R expresses the distance from O to S, θ is the angle of lines OS and OQ.

However, the electric potential in domain D' can not be derived from this method of electric images.

Example 3. Two Orthogonal Conducting Half-Planes As is shown in Fig. 9.3, a point electric charge Q (solid circle) is placed in the "corner" of two grounded orthogonal conducting half-planes with a distance a from the half-plane S_1 and b from the half-plane S_2. The problem is to find the electric potential in the "corner" space [4].

Solution

In this case, the boundary conditions of the original problem are the electric potential on the surfaces of the two orthogonal conducting half-planes S_1 and S_2 is zero, $\varphi = 0$ (on S_1 and S_2).

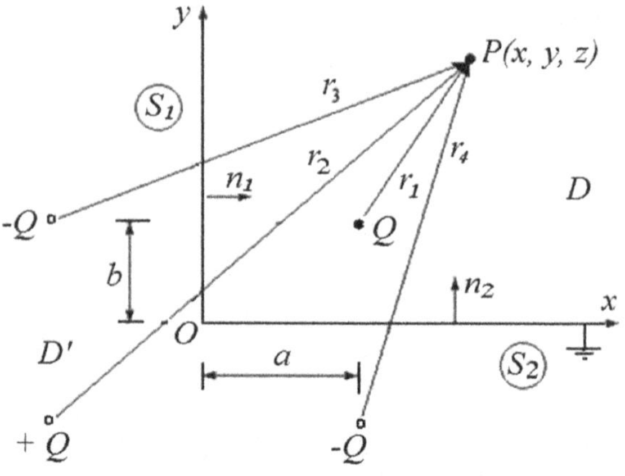

Fig. 9.3 An electric charge and two orthogonal conducting half-planes

The symmetric properties of the problem result in three hypothetical electric charges (hollow circles), which are located in domain D' in proper positions, respectively [4], Fig. 9.3.

As a result, the electric potential of an arbitrary $P(x, y, z)$ in domain D can be written as [4],

$$\varphi = \frac{Q}{4\pi\varepsilon_0}\left(\frac{1}{r_1} + \frac{1}{r_2} - \frac{1}{r_3} - \frac{1}{r_4}\right). \tag{9.22}$$

The density of induced charge on conducting plane S_1 is [4],

$$\sigma_1 = -\frac{Qa}{2\pi}\left(\frac{1}{r_1^3} - \frac{1}{r_2^3}\right). \tag{9.23}$$

The partial induced charge on the half-plane S_1 is

$$q_1 = \int_{S_1} \sigma_1 ds = -\frac{2Q}{\pi}\arctan\left(\frac{b}{a}\right). \tag{9.24}$$

Similarly, the partial induced charge on the half-plane S_2 is

$$q_2 = \int_{S_2} \sigma_2 ds = -\frac{2Q}{\pi}\arctan\left(\frac{a}{b}\right). \tag{9.25}$$

The total induced charges on the two half-planes is,

$$q = q_1 + q_2 = -\frac{2Q}{\pi}\left(\arctan\left(\frac{a}{b}\right) + \arctan\left(\frac{a}{b}\right)\right) = -Q. \tag{9.26}$$

It can be seen that the main feature of method of "electric images" is that the replacement of the real induced charges on the conductors by virtual or hypothetical electric charges outside the interested domain in the system as long as the boundary condition unchanged and met.

This is a miraculous approach that was used in electrostatics [4–6], which is expected to be used in engineering practices to solve complicated problem in complex systems.

9.5 Application of Nonlocal Euler–Bernoulli Beam Theory in Static Microstructure

Nonlocal elasticity was proposed by Eringen to take into account the scale effect in elasticity in continuum [7, 8], which assumes that the stress at a considered point is a functional of the strain field all over the body [9, 10]. In this way, the internal size scale could be involved into the constitutive equations simply by using a characteristic parameter of the material. The application of nonlocal elasticity, such as the nonlocal Euler–Bernoulli beam theory, has received much attention in micro- and nano-devices recently due to the rapid development of nanotechnology. The application of the nonlocal Euler–Bernoulli beam theory to micro and nano-devices has potentials to reveal scale effects for micro- and nano-devices [9, 10]. The aim of nonlocal theory is to consider "long-range interaction" of microstructures, it got achievements in interpreting wave propagation, diffraction of electromagnetic wave, polarization of light wave, dispersion phenomenon, etc. However, these phenomena could not be well explained by using classical local theory (local elasticity, etc.). Undoubtedly, the intrinsic characteristic of nonlocal theory is to consider the collective actions of material elements within the body, it is actually a "systemic view" in a sense. Here in this section an application of Euler–Bernoulli beam theory in static microstructure is presented.

The Euler–Bernoulli beam theory with nonlocal constitutive relation can be written as [9],

$$M - (e_0 a)^2 \frac{\partial^2 M}{\partial x^2} = -EI \frac{\partial^2 w}{\partial x^2}, \tag{9.27}$$

where E is the Young's modulus of the material; x is the coordinate with the origin at the left end of one-dimensional structure; $M(x, t)$ is bending moment on the beam; EI is the bending rigidity of beam structure; Parameter a describes *internal characteristic length of the material*; e_0 is a non-dimensional nonlocal scaling parameter which has been assumed to be a constant appropriate for each material; $w(x)$ is the static deformation of the structure.

The nonlocal Euler–Bernoulli beam theory can be further written alternately as [9],

$$EI\frac{d^4w(x)}{dx^4} - q(x)\left[1 - (e_oa)^2\frac{d^2q(x)}{dx^2}\right] = 0, \tag{9.28}$$

where $q(x)$ is the general external loading.

It can be easily seen from the Eq. (9.28) that when the parameter e_0 is set to be zero it reduces to the local Euler–Bernoulli beam theory.

1. **Nonlocal Euler–Bernoulli cantilever subjected to uniformly distributed force *f***

The solution for the nonlocal Euler–Bernoulli cantilever subjected to uniformly distributed force f is derived [9],

$$w(x) = \frac{f}{EI}\left[\frac{x^4}{24} - \frac{Lx^3}{6} + \frac{L^2x^2}{4} - \frac{x^2}{2}(e_0a)^2\right], \tag{9.29}$$

where L is the length of the cantilever.

Equation (9.29) indicates that the deformation at every point of structure is with the scale effect surely.

Obviously, it can be seen from Eq. (9.29) that the solution for the cantilever subjected to uniformly distributed force f loading of the local Euler–Bernoulli beam theory can be obtained easily by setting the scale coefficient e_0a to be zero.

Furthermore, the nonlocal Euler–Bernoulli cantilever could be numerically studied to show the scale effect on the static deformation of micro- and nano-bars, which is represented in Fig. 9.4. In the studies, the expressions of non-dimensional parameters, $\psi = EIw(x)/fL^4$, $\xi = x/L$, and $\lambda = e_0a/L$, are employed for structures subjected uniformly distributed force, respectively. Besides, the non-dimensional λ denotes the scale coefficient.

Thus an alternative form of Eq. (9.29) is obtained,

$$\psi = \frac{\xi^4}{24} - \frac{\xi^3}{6} + \frac{\xi^2}{4} - \frac{\xi^2}{2}\lambda^2. \tag{9.30}$$

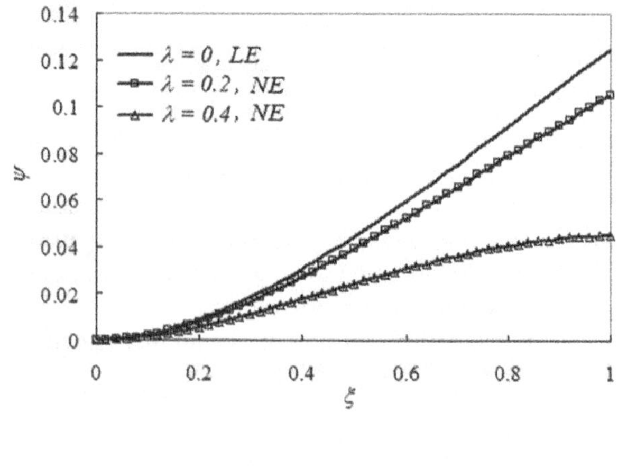

Fig. 9.4 Scale effect of the nonlocal Euler–Bernoulli cantilever subjected to uniformly distributed force on its static deformation

Moreover, the symbols for solutions for *local* and *nonlocal Euler–Bernoulli cantilevers* are by *LE* and *NE* comparatively, respectively. Especially, $\lambda = 0$ is corresponding to the local Euler–Bernoulli cantilever, *LE*.

Figure 9.4 shows obvious difference between *local Euler–Bernoulli cantilever* and *nonlocal Euler–Bernoulli cantilever*, which indicates the scale effect on the static deformation of micro- and nano-bars.

2. **Nonlocal Euler–Bernoulli cantilever subjected to a concentrated force *P* at its free end**

The nonlocal Euler–Bernoulli cantilever subjected to a concentrated force P at its free end $x = L$, its solution is derived as [9],

$$w(x) = \frac{P}{EI}\left(\frac{Lx^2}{2} - \frac{x^3}{6}\right). \tag{9.31}$$

From Eq. (9.31), it can be seen that the nonlocal Euler–Bernoulli cantilever subjected to a concentrated force P at its free end $x = L$ does not induce the scale effect on the static deformation of the cantilever accidentally, i.e., under condition of a concentrated force P applied at its free end $x = L$ the solution for the cantilever of the nonlocal Euler–Bernoulli beam theory is the same as that of the local Euler–Bernoulli cantilever luckily.

Analogically, an alternatively dimensionless form of Eq. (9.31) can be obtained by denoting $\varphi = EIw(x)/PL^3$, $\xi = x/L$, and $\lambda = e_0a/L$,

$$\varphi = \frac{\xi^2}{2} - \frac{\xi^3}{6}. \tag{9.32}$$

However, the nonlocal Timoshenko cantilever gives a scale effect due to its shear action term under the same condition, which is written as [9],

$$w_T(x) = \frac{P}{EI}\left(\frac{Lx^2}{2} - \frac{x^3}{6}\right) + \frac{(1+\nu)PL^2}{8\kappa EI}\left(\frac{d}{L}\right)^2 x. \tag{9.33}$$

In Eq. (9.33), ν is Poisson's ratio, d is the diameter of the rod or tube in the study; parameter κ is the recommended adjustment coefficient, which takes 10/9 for a circular shape of the cross area [9]. The last term in Eq. (9.33), $\frac{(1+\nu)PL^2}{8\kappa EI}\left(\frac{d}{L}\right)^2 x$, reflects the additional action of nonlocal Timoshenko cantilever under condition of a concentrated force P at its free end comparatively. The coefficient, L/d, represents the slenderness of the structure [9].

The nonlocal Euler–Bernoulli cantilever could be also numerically studied, which is represented in Fig. 9.5.

3. **Simply supported nonlocal Euler–Bernoulli beam subjected to pure bending**

Simply supported nonlocal Euler–Bernoulli beam subjected to pure bending with bending moment M, its solution is derived as [9],

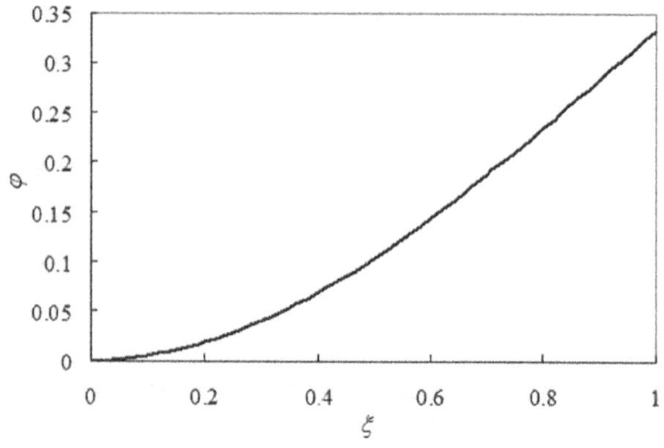

Fig. 9.5 The nonlocal Euler–Bernoulli cantilever subjected to a concentrated force P at its free end

$$w(x) = \frac{M}{EI}\left[\frac{Lx}{2} - \frac{x^2}{2} + (e_0a)^2\right]. \tag{9.34}$$

Similarly, another dimensionless form of Eq. (9.34) can be obtained by denoting $\eta = EIw(x)/ML^2$, $\xi = x/L$, and $\lambda = e_0a/L$ alternatively,

$$\eta = \frac{1}{2}(\xi - \xi^2 + \lambda^2). \tag{9.35}$$

Obviously, Eq. (9.35) indicates that the nonlocal effect in this case is invariable quantity, $\lambda^2/2$, in range of $0 \leq x \leq L$. It can be easily seen from above two equations that the solutions of nonlocal beam theory can reduce to that of local beam theory if e_0a or λ is zero.

The scale effect of simply supported nonlocal Euler–Bernoulli beam subjected to pure bending on the static deformation could be also numerically investigated, which is represented in Fig. 9.6.

9.6 Summary

Through above discussions, it could see that approaches in systemic view can be found in various real fields broadly, ranging from "five essential elements" system of ancient China, to modern science, including ecology system, electrostatics, and nonlocal theory, which again indicates the universality of systemic phenomena and the relevant theoretical methods.

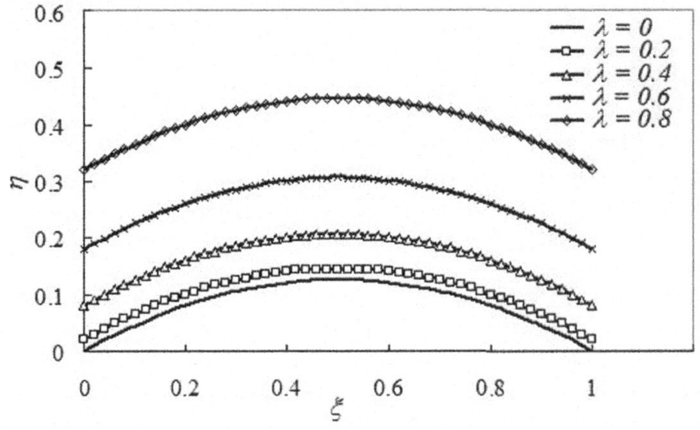

Fig. 9.6 Scale effect of simply supported nonlocal Euler–Bernoulli beam subjected to pure bending on the static deformation

References

1. W. Lin, *Ecology*, 2nd edn. (Science Press, Beijing, China, 2024)
2. Y. Lin, R. Fu, *Ecology*, 3rd edn. (Science Press, Beijing, China, 2023)
3. M. Zheng, J. Yu, *Robust Design and Assessment of Product and Production by Means of Probabilistic Multi-objective Optimization* (Springer, Singapore, 2024)
4. M. Chaichian, I. Merches, D. Radu, A. Tureanu, *Electrodynamics, An Intensive Course*, (Springer, Heidelberg, 2016)
5. K. Bhattacharya, S. Mukhopadhyay, *Introduction to Advanced Electrodynamics* (Springer, Singapore, 2021)
6. S. Guo, Q. Huang, Z. Li, Q. Lin, *Electrodynamics*, 4th edn. (Higher Education Press, Beijing, 2023)
7. A.C. Eringen, D.G.B. Edelen, On nonlocal elasticity. Int. J. Eng. Sci. **10**, 233–248 (1972)
8. A.C. Eringen, On differential equations of nonlocal elasticity and solutions of screw dislocation and surface waves. J. Appl. Phys. **54**(9), 4703–4710 (1983)
9. Q. Wang, K.M. Liew, Application of nonlocal continuum mechanics to static analysis of micro- and nano-structures. Phys. Lett. A **363**, 236–242 (2007)
10. J. Chen, *Nonlocal Euler–Bernoulli Beam Theories, A Comparative Study*, (Springer, Cham, Switzerland, 2021)

Chapter 10
General Conclusions

Abstract This chapter gives a summary of the whole book, mainly the intrinsic essence of systems theory for engineering practice in sight of physics.

Keywords Systems theory · Physics · Cooperative effects · Synergetics · Wholeness · Optimization

Through thorough descriptions in this book, the authors would like to summarize following terms comprehensively:

1. The wholeness, irreplaceability of components, and organic integral of a system are main characteristics of systems theory;
2. The optimization of a system is its integral/overall optimization as a whole;
3. The typical problems in classical vibration and wave in physics all reveal group behavior of the system by using proper models;
4. The extreme value problems in natural world exhibit group behavior of the system and the synergetic actions among all parts in either spatial or temporal scales;
5. The states of electrons in atom and metals reflect the group and collective behavior of the electronic system and their synergetic actions among all parts of participation;
6. Ising model in phase transition, elementary excitation and percolation approach in solid all reveal cooperative effects, collective phenomena and wholeness in the system;
7. The homogeneity and ergodicity of constituting components with specific requirements of a composite system in macroscale grossly in statistical physics are indispensable in effective media approximation;
8. "Unity of opposites" is the basic rule of universe inside a system in aspects of both large spatial and temporal scales, which determines the overall certainty from individual uncertain phenomena;
9. The involved methodologies for systems theory are organic combination and synergism of both "reductionism" and "holism" instead of "confrontation" or "separation" of them, which could be used to deal with analogous problems

M. Zheng and J. Yu, *Systems Theory for Engineering Practice*,
https://doi.org/10.1007/978-981-97-9342-6_10

in systems science-engineering fields and to promote the transformation of wisdom to productivity;

10. Further exploration of systemic thinking in extensive fields is expected, which is with bright prospects. The living soul of Marxism is "seeking truth from facts", "abandoning dogma" and "concrete analysis". One should proceed from reality, constantly raise awareness, explore and reveal laws, so as to approach to the truth.

SPRINGER NATURE

GPSR Compliance

The European Union's (EU) General Product Safety Regulation (GPSR) is a set of rules that requires consumer products to be safe and our obligations to ensure this.

If you have any concerns about our products, you can contact us on ProductSafety@springernature.com

In case Publisher is established outside the EU, the EU authorized representative is:

Springer Nature Customer Service Center GmbH
Europaplatz 3
69115 Heidelberg, Germany

The manufacturer's authorised representative in the EU is Springer Nature Customer Service Centre GmbH, Europaplatz 3, 69115 Heidelberg, Germany. If you have any concerns regarding our products, please contact ProductSafety@springernature.com

Printed and bound by CPI Group (UK) Ltd, Croydon, CR0 4YY
01/10/2026
02225875-0012